Algebra
for the
Utterly Confused

Algebra
for the
Utterly Confused

Larry J. Stephens

McGraw-Hill

New York   San Francisco   Washington, D.C.   Auckland   Bogotá
Caracas   Lisbon   London   Madrid   Mexico City   Milan
Montreal   New Delhi   San Juan   Singapore
Sydney   Tokyo   Toronto

Library of Congress Cataloging-in-Publication Data applied for.

 2 3 4 5 6 7 8 9 10 11 12 13 14 15 DOC/DOC 0 9 8 7 6 5 4 3 2 1

ISBN 0-07-135514-6

MINITAB™ is a trademark of Minitab Inc., and is used herein with the owner's permission.

The images used herein, by permission, were obtained from IMSI's Master Clips and
Master Photos Premium Image Collection, 75 Rowland Way, Novato, CA 94945

*The sponsoring editor for this book was Barbara Gilson, the editing supervisor
was Maureen B. Walker, and the production supervisor was Tina Cameron. It
was set in Times Ten by North Market Street Graphics.*

Printed and bound by R. R. Donnelley & Sons Company.

McGraw-Hill books are available at special quantity discounts to use as
premiums and sales promotions, or for use in corporate training programs. For
more information, please write to the Director of Special Sales, McGraw-Hill,
Two Penn Plaza, New York, NY 10121-2298. Or contact your local bookstore.

 This book is printed on recycled, acid-free paper containing a minimum
of 50% recycled, de-inked fiber.

Contents

A Special Message to the Utterly Confused Algebra Student

What qualifications are needed to write a book that will help the utterly confused algebra student? I believe that I can answer that question. Certainly having a B.S. in Mathematics and Physics, an M.A. in Mathematics, and a Ph.D. in Statistics, as I do, helps. Having taught mathematics and statistics for over 30 years, as I have, also helps. However, having been confused about algebra a long time ago myself helps tremendously in writing such a book. Prior to my senior year in high school, I was utterly confused about algebra. In fact my grades in first-year algebra, taken in my freshman year were all C or below. I successfully avoided math courses in my sophomore and junior years. However, in my senior year I made all As in second-year algebra, plane geometry, and in a course called solid geometry. Sometime before my senior year, the Russians put a small satellite called *Sputnik* in orbit. This event started the space race between Russia and the United States. Space explorations changed my *attitude* toward mathematics and science.

A desire to learn algebra and a willingness to spend the time necessary to fully comprehend the concepts are extremely important in overcoming your confusion about algebra. I do not pretend to be able to give everyone a *Sputnik* experience. Unfortunately, there is no substitute for the hours needed to practice the many and varied concepts that make up algebra. I have worked for NASA, Los Alamos Scientific Laboratory, Lawrence Radiation Laboratory, 3M, and taught at the college level for over 30 years. Although the mathematics I used in my jobs was often beyond that found in algebra, it was the foundation in algebra that supported the advanced mathematical techniques that were used. Algebra is the foundation of all the math courses that follow as well as many of the courses in science, business, and other areas of study. A good understanding of algebra is extremely important to your future success in college and to your career.

Algebra for the Utterly Confused explains the concepts of algebra by the use of practical examples whenever possible. The power of the computer has also been used in trying to increase your understanding of algebra. Output from powerful software packages such as MAPLE, Minitab, and Excel are integrated throughout the book to help you understand the concepts of algebra. A set of icons are used throughout each chapter to help guide students in their study of the book. The icons are described on the following two pages.

My goal in *Algebra for the Utterly Confused* is to clarify topics in algebra that have traditionally confused students.

Icons

Don't Forget

This icon highlights things you should memorize. Right before a test, go over these items to keep them fresh in your mind.

Quick Tip

This icon appears next to the "deeper" insights into a problem. If you have trouble understanding the details of why a problem makes physical sense, then this is the icon to follow.

Danger!

This icon highlights trouble spots and common traps that students encounter. If you are worried about making frustrating little mistakes or think you are losing points on tests because of missing little "tricks" then this is the icon to follow.

Pattern

 The intention of this icon is to help identify a pattern of solving one problem that works for a general category of problems. In many cases the pattern is reviewed in a step-by-step summary along with examples of similar problems.

Short Cuts

 Items next to this icon can be skipped if you are really struggling. On a second pass through the book, or for the more advanced student, this icon is intended to show a few extra tricks that will allow you to do problems faster. These items are included because speed is often important to success on algebra tests.

Test Yourself

 This icon indicates a set of problems that allow you to test yourself to see if you understand the material presented in the chapter.

How to Study Algebra

Being successful in your effort to learn algebra is no different than learning to play the piano, being successful at playing golf, or many of the other human endeavors that you attempt. You must be willing to *practice the techniques* that are presented to you. Remember:

1. *Algebra is not learned by osmosis.* You will not automatically absorb algebra by simply attending class. You must work a lot of problems to fully understand the concepts.

2. *Algebra is not a spectator sport.* You must be an active participant in the learning process.

The topics and concepts in algebra are *sequentially dependent.* This means that the topic you are studying today is dependent on topics you learned yesterday, and the topics you study tomorrow will depend on the ones you learn today. For example, the distributive law states that for any numbers a, b, and c, that $a(b + c) = ab + ac$. The distributive law is usually learned early in algebra. But it is applied again and again in topics such as factoring and simplifying algebraic expressions.

There are at least two aspects always present in algebra. These are the *mechanical* and the *conceptual* aspects. These two aspects are not mutually exclusive. That is, both may be present. When finding the roots of a quadratic equation, a sequence of mechanical steps is applied to determine the roots. However, a *story problem* may lead to a quadratic equation that must be solved in order to obtain the solution to the problem. A conceptual understanding of the story problem is needed in order to set up the problem. Students usually have more difficulty with story problems than with the mechanical steps used to solve a problem. To be successful at both mechanical and conceptual problems, work as many problems as possible of both types.

General Guidelines for Effective Algebra Study

1. Starting on the very first day of classes, systematically work problems of all types until you are confident that you understand all concepts.

2. Be sure to read the discussion given in the text of the sections covered on a given day. Work your way through all examples in the text. Have a pencil and paper close by and fill in any missing details in the examples. If there are parts of the examples that you do not understand, ask your instructor to help you fill in the details.

3. Do not get behind in the class. Once you get behind in the class, the *snowball effect* follows. The new concepts that you encounter usually depend on your understanding of the ones you are behind on.

4. Take advantage of any sources of help such as tutoring labs, your instructor's office hours, and any extra credit possibilities. I recently taught algebra and offered extra credit for some computer algebra assignments to be done using the computer package MAPLE. I was surprised at the large number of students who did not take advantage of the extra credit opportunities.

5. If you like to work with others, try to select some good students to work with from time to time.

6. If you are struggling with your algebra course, consider some supplemental materials such as *Algebra for the Utterly Confused*.

Preparing for Tests

1. Your best test preparation is to keep up in the course daily. If you do this, you will not need to stay up late the night before the test.

2. Get a good night's sleep the night before the test. You are much more likely to be able to think clearly and make good decisions if you are rested when you take the test. **Never stay up all night studying the night before an algebra test.**

3. I have always found it useful to take a simulated test before taking the real test. Go through the sections that the test covers and select some problems similar to those the test covers that you have not worked before. Choose ones that have answers. If there are not enough in your text, choose some from another algebra book from the library. After selecting these problems, form a simulated test and then work this test just as you would the real one you will take tomorrow.

Test-Taking Strategies

1. If you are having difficulty with a given problem, leave it and go to another one that is easier. Save the harder problems until you have worked all the easy ones.

2. If you finish the test early, be sure to check over your work. I am surprised at the number of students in my algebra courses who still rush through the test and turn it in early without checking it over.

3. Do not hesitate to ask for clarification on any problems that are not clear to you.

4. If you are allowed to use a calculator on the test, be sure to use it to your maximum advantage. Sometimes by substituting some simple numbers in an expression you may be able to determine the reasonableness of a solution.

Preface

Algebra is among the most difficult courses for many high school and college students to master. The sequential steps involved in solving algebra problems require repetition for their mastery. Many students do not perform the repetitions needed to master the concepts in algebra.

Another factor that makes algebra difficult is the abstract nature of the course. Students fail to see the relevance of algebra to the real world. Even the algebra books that have attempted to incorporate more real-world problems have not always succeeded in capturing the student's interest in algebra. I believe students can be motivated to study algebra if they can see the relevance of algebra.

Algebra for the Utterly Confused is not intended to be a textbook in algebra. There are numerous textbooks in algebra and many of them are very well written. *Algebra for the Utterly Confused* is intended to be used as a supplement to algebra courses in high school as well as in college. I have covered many, but not all, topics found in such courses. All algebra textbooks have many examples and numerous problems. I have included a section at the end of each chapter entitled Test Yourself. This section contains problems similar to the concepts discussed in the chapter and will allow you to determine whether you understand the material. The examples I have included are intended to be current and relevant to the real world. It is not possible to relate every algebra topic to the real world. For example, it is difficult to connect the imaginary unit or complex numbers to applications in the real world. However, I do mention that they are useful in advanced areas such as electrical engineering.

Another aspect of the book not found in algebra textbooks is the inclusion of computer software that is relevant to algebra. In a 1999 article by Dr. John Konvalina and me in the *International Journal of Mathematical Education in Science and Technology* entitled "The Use of Computer Algebra Software in Teaching Intermediate and College Algebra" (Vol. 30, No. 4, pp. 483–488), we discuss the effect of using computer algebra software in the teaching of algebra. We found that the use of such software increased the mean performance on a common final exam when a class using the software was compared to a class not using it. In addition, the students had very positive attitudes toward the incorporation of the software in the course, and the authors of the article received their best student evaluations ever when teaching algebra courses.

I have chosen to include output from three different software programs. MAPLE (Waterloo Maple Inc., 57 Erb Street West, Waterloo, Ontario, Canada, N2L 6C2) is a computer algebra software package and it was used in the study mentioned in the preceding paragraph. MINITAB™ (Minitab Inc., 3081 Enterprise Drive, State College, PA. 16801-3008) is a software package having statistical as well as excellent graphing and spreadsheet capabilities. MINITAB is a trademark of Minitab Inc. and is used herein with the owner's permission. Excel spreadsheet is widely available as a segment of Microsoft Office that many people have as part of their personal computing software. I wish to thank Waterloo Maple Inc., Minitab Inc., and Microsoft Corporation for permission to include output from their software. The inclusion of this software output helps enrich the study of algebra as well as indicate its relevancy to the real world.

The algebra topics have been "spiced up" by the use of computer clip art throughout the book. I wish to thank IMSI (75 Rowland

Way, Novato, CA 94945) for permission to include clip art images from Master Clips in the book.

Throughout the book, topics and concepts are discussed in the setting of some real-world examples. Usually a scenario is described along with a problem related to that scenario. As the problem and its solution are discussed, algebra terms and concepts are defined, discussed, and described.

The purchaser of this book is encouraged to use it along with whatever algebra textbook you may be using in your current algebra course. When a difficult section is encountered, turn to *Algebra for the Utterly Confused* to get you over whatever difficulty you have encountered. Or if you are not currently in an algebra course, but are using some algebra-related concepts in your job and need a refresher for algebra, turn to *Algebra for the Utterly Confused* to review algebra for you.

In addition to thanking Minitab Inc., Waterloo Maple Inc., Microsoft Corporation, and IMSI, I would like to thank the editor, Barbara Gilson, and the staff at McGraw-Hill for their help. I would also like to thank my good friend of over 25 years, Dr. John Konvalina, Mathematics Department, University of Nebraska at Omaha for our many discussions concerning mathematics. We have also been joint authors on many mathematics education research publications. Finally, I would like to thank my wife Lana for her support and many helpful suggestions concerning the book.

—Dr. Larry J. Stephens
Omaha, Nebraska

CHAPTER 1

Basic Algebra Concepts Review

Do I Need to Read This Chapter?

➡ What is the real number system?

➡ How do I add, subtract, multiply, and divide fractions?

➡ How do I add, subtract, multiply, and divide signed numbers?

➡ What is scientific and exponential notation?

➡ What is a prime factorization?

➡ What are algebraic variables?

➡ How do I perform operations on algebraic expressions?

➡ How do I factor algebraic expressions?

➡ How do I solve equations?

➡ What is computer algebra software?

➡ What is MAPLE?

Real Numbers System

Example 1-1: Table 1-1 gives a portion of the stock market page for a Wednesday and a Thursday similar to that found in the Money section of *USA Today.* The table gives the highest and lowest price per share during the past year, the name of the stock, close (the price per share at the end of the day), and the change from the day before. Several different types of numbers appear in the table. Positive and negative whole numbers and positive and negative fractions are illustrated in the table.

Table 1-1 Stock Market Information

		WEDNESDAY		
52-WEEK HIGH	**52-WEEK LOW**	**STOCK**	**CLOSE**	**CHANGE**
$96\frac{1}{8}$	$48\frac{3}{8}$	AT&T	$84\frac{9}{16}$	$-1\frac{3}{8}$
$77\frac{5}{16}$	$57\frac{9}{16}$	Exxon	$67\frac{55}{64}$	$+\frac{7}{64}$
105	69	GE	101	-2
$199\frac{1}{4}$	$95\frac{7}{8}$	IBM	$173\frac{3}{4}$	$-3\frac{3}{16}$
$134\frac{7}{8}$	$84\frac{1}{8}$	Pfizer	$132\frac{1}{2}$	$-1\frac{3}{8}$
		THURSDAY		
52-WEEK HIGH	**52-WEEK LOW**	**STOCK**	**CLOSE**	**CHANGE**
$96\frac{1}{8}$	$48\frac{3}{8}$	AT&T	$84\frac{1}{2}$	$-\frac{1}{16}$
$77\frac{5}{16}$	$57\frac{9}{16}$	Exxon	$67\frac{55}{64}$	—
105	69	GE	$100\frac{3}{8}$	$-\frac{5}{8}$
$199\frac{1}{4}$	$95\frac{7}{8}$	IBM	$173\frac{5}{8}$	$-\frac{1}{8}$
$134\frac{7}{8}$	$84\frac{1}{8}$	Pfizer	$130\frac{7}{8}$	$-1\frac{5}{8}$

The real numbers are represented visually by the **real number line.** This representation is shown in Fig. 1-1. The line is shown as stretching from minus infinity to plus infinity. That is, from extremely small negative numbers to extremely large positive numbers. If the number is positive, it is measured to the right of zero. If the number is negative, it is measured to the left of zero. Every real number is represented by some point on this line. Conversely, each point on this line represents some real number.

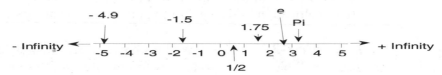

Fig. 1-1

A few particular points are shown in the figure. The real numbers are either **rational numbers** or **irrational numbers.** The positive and negative whole numbers and fractions are rational numbers. There are other numbers that are not whole numbers or fractions. They are called irrational numbers. The number π (pi) is an irrational number. It is not equal to a fraction. In Fig. 1-1, π (pi) is shown between 3.1 and 3.2. Another irrational number is the number c, which is encountered in calculus. It is shown in Fig. 1-1 between 2.71 and 2.72. All the numbers in Table 1-1 are located somewhere on this real number line. Any fraction has a decimal equivalent.

For example, $\frac{1}{2}$ is equivalent to 0.5. The decimal equivalent is obtained by dividing the fraction. The decimal equivalent for $\frac{1}{3}$ is 0.333 · · · . The decimal equivalent of $\frac{1}{7}$ is 0.142857142857 · · · . A rational number will have a decimal equivalent that either *terminates,* such as $\frac{1}{2} = 0.5$ and 3/4 = 0.75, or *repeats* itself such as 1/3 = 0.333 · · · and $\frac{1}{7} = 0.142857\,142857$ · · · . Irrational numbers neither terminate nor repeat. Pi and e do not terminate and they do not have a repeating pattern.

Four Fundamental Operations Involving Fractions

The four fundamental operations performed on real numbers are addition, subtraction, multiplication, and division. When fractions and/or signed numbers are involved, special rules and techniques are developed in basic algebra to perform the four operations.

Example 1-2: Using the Wednesday and Thursday closing values given in Table 1-1, illustrate how to operate on fractions to obtain the change value given on Thursday for a share of AT&T.

The change for Thursday is equal to Thursday's close, $84\frac{1}{2}$, minus Wednesday's close, $84\frac{9}{16}$. Subtracting the whole number parts, we obtain $84 - 84 = 0$. Subtracting the fractional parts we have $\frac{1}{2} - \frac{9}{16}$. To perform the subtraction, express $\frac{1}{2}$ as $\frac{8}{16}$ and subtract the numerators and put the result over the common denominator as follows.

$$\frac{8}{16} - \frac{9}{16} = -\frac{1}{16}$$

Example 1-3: Suppose the IBM closing value in Table 1-1 was split into halves on Wednesday. Using the Wednesday closing value for IBM, what would each new share be worth?

To find the new value of a share, we need to take half the per share value, $173\frac{3}{4}$. This is accomplished as follows.

Step 1: $\quad \frac{1}{2}\left(173\frac{3}{4}\right) = \frac{1}{2}\left(\frac{695}{4}\right)$

Step 2: $\quad \frac{1}{2}\left(\frac{695}{4}\right) = \frac{695}{8}$

Step 3: $\quad \frac{695}{8} = 86\frac{7}{8}$

In step 1, the mixed number $173\frac{3}{4}$ is converted to the improper fraction $\frac{695}{4}$ by multiplying 4 times 173 and adding 3 and putting the result over 4. In step 2, the numerators are multiplied and the result is put over the product of the denominators. In step 3, the improper fraction $\frac{695}{8}$ is converted to the mixed number $86\frac{7}{8}$ by noting that 8 divides into 695 eighty-six times with remainder 7, which is put over 8 to form the fraction $\frac{7}{8}$. Each new share will be worth \$$86\frac{7}{8}$ after the split occurs.

Example 1-4: Find the percent change for Pfizer on Thursday. The percent change for Pfizer on Thursday is the Thursday change divided by the Wednesday close and multiplied by 100 to give a percent. The steps are as follows.

Step 1: $\quad -1\frac{5}{8} \div 132\frac{1}{2} = -\frac{13}{8} \div \frac{265}{2}$

Step 2: $\quad -\frac{13}{8}\left(\frac{2}{265}\right) = \frac{26}{2120}$

Step 3: $\quad -\left(\frac{26}{2120}\right)(100) = -\frac{2600}{2120} = -1.23\%$

In step 1, the Thursday change and the Wednesday close for Pfizer are copied from Table 1-1 and changed from mixed form to improper fraction form. In step 2, we invert the second fraction and then multiply the resulting fractions. In step 3, we convert the fraction to a decimal and multiply by 100 to find the percent change. A share of Pfizer dropped 1.23% on Thursday.

Pattern

1. To add or subtract fractions, first obtain a common denominator and add or subtract the numerators and place the result over the common denominator as illustrated in *Example 1-2*.

2. To multiply fractions, multiply the numerators and put the result over the product of the denominators.

3. To divide fractions, invert the second and then multiply.

Short Cuts

Often, arithmetic with fractions may be speeded up by converting the fractions to their decimal equivalents and then performing the operations. The next examples illustrate how to do this.

Example 1-5: Work **Example 1-2** using decimal representation rather than fractional representation.

Thursday's close was $84\frac{1}{2} = 84.5$ and Wednesday's was $84\frac{9}{16} = 84.5625$ and the difference is $84.5 - 84.5625 = -0.0625$. The decimal number -0.0625 is equal to the fraction $-\frac{625}{10,000}$. If this fraction is reduced by dividing both the numerator and the denominator by 625 the result is $-\frac{1}{16}$.

Example 1-6: Work **Example 1-3** using decimal representation rather than fractional representation.

The cost per share is $173\frac{3}{4} = 173.75$. When the shares split, the new cost per share will be $\frac{173.75}{2} = 86.875$. The decimal number 86.875 is equal to 86 plus the fraction $\frac{875}{1000}$. If $\frac{875}{1000}$ is reduced by dividing numerator and denominator by 125, the result is $\frac{7}{8}$. Thus the cost per share after splitting is $86\frac{7}{8}$.

Example 1-7: Work **Example 1-4** using decimal representation rather than fractional representation.

The percent change for Pfizer on Thursday is the change given on Thursday divided by the Wednesday close and multiplied by 100 to give a percent. The change Thursday was $-1\frac{5}{8} = -1.625$ and the Wednesday close was $132\frac{1}{2} = 132.5$. The percent change is given by $(-1.625/132.5) \times 100 = -1.23\%$.

Four Fundamental Operations Involving Signed Numbers

The changes in stock prices in Table 1-1 are positive for stocks that increased in value and negative for stocks that decreased in value. The sign of the change indicates whether the stock price rose or fell. In order to trace the behavior of a stock, an understanding of operations involving signed numbers is necessary.

Example 1-8: Suppose the daily changes for IBM stock were –2, +3, +1, –4, and –1. What was the change for the week? The change for the week was –2 +3 +1 –4 –1 = –3. For the week, the stock was down 3 points.

Example 1-9: Suppose IBM stock were down 2 points each day for the week. How many points is IBM down for the week?

$$-2 -2 -2 -2 -2 = 5(-2) = -10$$

The IBM stock is down 10 points for the week. Note that a positive 5 times a negative 2 gives a negative 10.

Example 1-10: Using the results of **Example 1-9,** we see that $-10 \div -2 = 5$. This is obtained from the result $5(-2) = -10$ by dividing both sides by -2. This is illustrated as follows.

$$\frac{-10}{-2} = \frac{5(-2)}{(-2)} = 5$$

In general, dividing a negative number by a negative number gives a positive result.

An important property of signed numbers is their **absolute value.** The absolute value of a number is the numerical value of the number exclusive of its sign. The absolute value of $+5$ is 5 and the absolute value of -2.5 is 2.5.

Pattern

1. The sum of two positive numbers is positive and the sum of two negative numbers is negative. For example, $5 + 7 = 12$ and $-5 + (-5) = -10$.

2. The sum of a positive and a negative number is positive, if the larger number (in absolute value) is positive, and the sum is negative if the larger number (in absolute value) is negative. The sum of a positive and a negative number having the same absolute value is zero. For example, $3 + (-5) = -2$, $9 + (-5) = 4$, and $5 + (-5) = 0$.

3. To subtract two signed numbers, change the sign of the second number and then add. For example, $8 - (-5) = 8 + 5 = 13$, $-5 - (-5) = -5 + 5 = 0$, and $13 - (+7) = 13 + (-7) = 6$.

4. The product of a positive number and a negative number is negative. The product of two positive numbers or the product of two negative numbers is positive. For example, $7(-5) = -35$, $(-5)(9) = -45$, and $(-5)(-35) = 175$.

5. The quotient of a positive number and a negative number is negative and the quotient of two negative numbers or two positive numbers is positive. For example, $-12/4 = -3$, $16/(-8) = -2$, and $(-12)/(-2) = 6$.

Scientific and Exponential Notation

The salaries of professional sports figures have become so large in recent years that it is now reasonable to express them in scientific notation. Kevin Brown of the Los Angeles Dodgers recently signed a 7-year contract for $105 million. If this is expressed in digits, it is equal to $ 105,000,000. It takes nine digits to completely express this amount.

Example 1-11: Suppose we wish to express Kevin Brown's contractual salary in scientific notation. First, we need to discuss **powers of 10.** Powers of 10 are illustrated in Table 1-2.

Table 1-2   Powers of 10

POWER OF TEN	MEANING	VALUE
10^1	10	10
10^2	10×10	100
10^3	$10 \times 10 \times 10$	1,000
10^4	$10 \times 10 \times 10 \times 10$	10,000
10^5	$10 \times 10 \times 10 \times 10 \times 10$	100,000
10^6	$10 \times 10 \times 10 \times 10 \times 10 \times 10$	1,000,000
10^7	$10 \times 10 \times 10 \times 10 \times 10 \times 10 \times 10$	10,000,000
10^8	$10 \times 10 \times 10 \times 10 \times 10 \times 10 \times 10 \times 10$	100,000,000
10^9	$10 \times 10 \times 10 \times 10 \times 10 \times 10 \times 10 \times 10 \times 10$	1,000,000,000

To express Kevin Brown's salary in scientific notation, we express $105 million as follows: $1.05 \times 100{,}000{,}000 = 1.05 \times 10^8$. In **scientific notation,** a number is expressed as the product of a number between 1 and 10 (including 1 but not 10), or -1 and -10 and 10 to some power. Notice that for Brown's salary, 1.05 is between 1 and 10 and the power of 10 is 8. In the power of 10, 10^8, 10 is called the **base** and 8 is called the **exponent.** Powers of 10 are a special case of exponential notation. If a is a real number and n is a natural number $(1, 2, 3, \cdots)$ then a^n means multiply a times itself n times. This is called **exponential notation** and a is referred to as the **base** and n as the **exponent.** Two important **laws of exponents** are as follows:

$$a^m \, a^n = a^{m+n} \qquad \text{and} \qquad a^m/a^n = a^{m-n}$$

Basically, these laws tell us that when you multiply two numbers in exponential notation having the same base you add exponents, and when you divide them you subtract exponents.

Example 1-12: Suppose we want to determine how much Kevin Brown makes per day for the 7 years in which he is paid $105 million. Assuming 365 days per year, there are 7 times $365 = 2555$ days in the 7 years. The amount per day is $105 million/2555 days. As we have seen, $105 million is equal to 1.05×10^8. The number of days express in scientific notation is equal 2.555×10^3. Dividing, we obtain the following.

$$\frac{1.05 \times 10^8}{2.555 \times 10^3} = 0.41096 \times 10^5$$

Note that when 10^8 is divided by 10^3, the exponents are subtracted to obtain 10^5. The division $1.05/2.555$ was accomplished by the use of a calculator. The pay per day is $41,096.

Danger!

A common mistake that students sometimes make when applying the laws of exponents occurs when different bases are involved. The following example illustrates this situation.

Example 1-13: Suppose we wish to multiply 2^2 by 4^3. Since $2^2 = 4$ and $4^3 = 64$, we know the answer is $(2^2)(4^3) = 4 \times 64 = 256$. In addition, we find $4^3/2^2 = 64/4 = 16$. **The following illustrates two common mistakes.**

$$(2^2)(4^3) \neq 8^{2+3} = 8^5 = 32{,}768$$

and

$$\frac{4^3}{2^2} \neq \left(\frac{4}{2}\right)^{3-2} = 2^1 = 2$$

In other words, the rules of exponents stated previously for multiplication and division work only when the bases are the same. Otherwise you have to find a different way, as we did at the beginning of the example.

Prime Factorization

A positive integer greater than 1 that is divisible only by 1 and itself is called a **prime number.** Some of the smaller prime numbers are 2, 3, 5, 7 and 11. Every positive integer can be expressed as a prime or a product of primes known as **prime factorization.** This concept introduces students to the algebra concept of factoring. A few examples serve to illustrate this concept.

Example 1-14: Suppose we wish to find the prime factorization of 128. We start by finding all factors of 2 in 128.

$$128 = 2(64) = 2(2)(32) = 2(2)(2)(16) = 2(2)(2)(2)(8) = 2(2)(2)(2)(2)(4)$$
$$128 = 2(2)(2)(2)(2)(2)(2)$$

So,

$$128 = 2^7$$

Example 1-15: Suppose we wish to find the prime factorization of 258. We start by finding all factors of 2 in 258. Then, we look for factors of 3. Once 43 is obtained, we stop because 43 is prime. We say that 258 is expressed as the product of the factors 2, 3, and 43.

$$258 = 2(129) = 2(3)(43)$$

Quick Tip

When factoring a positive integer into a product of primes, first factor all 2s out (if any), then factor all 3s out (if any), then factor all 5s out (if any), and continue in this fashion until all the factors are prime numbers.

Algebraic Variables

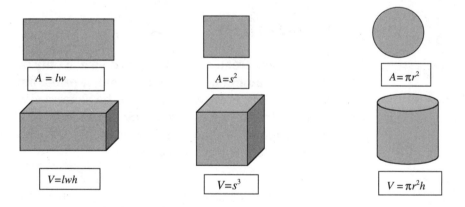

$A = lw$

$A = s^2$

$A = \pi r^2$

$V = lwh$

$V = s^3$

$V = \pi r^2 h$

A **variable** is a letter that is used to represent a number. A rectangle has area equal to length times width. If A represents the area, l represents the length, and w represents the width, then $A = lw$. In the formula for the area of a rectangle, A, l, and w are variables. Similarly, the area of a square is $A = s^2$, where s is the length of a side. The area of a circle is $A = \pi r^2$, where r is the radius of the circle. A rectangular box having dimensions l, w, and h has volume $V = lwh$. The volume of a cube is $V = s^3$. The volume of a cylinder is equal to $V = \pi r^2 h$. These six formulas illustrate how physical dimensions such as length, area, and volume may be represented by formulas involving variables.

Operations with Algebraic Expressions

Example 1-16: A house has two upstairs bedrooms with dimensions x by $x + 2$, three downstairs bedrooms with dimensions $x + 2$ by $x + 4$, a kitchen and a dining room each having dimensions $x + 2$ by $x + 2$. What is the total area of these seven rooms?

The area of an upstairs bedroom is $x(x + 2)$ and there are two of these. The area of a downstairs bedroom is $(x + 2)(x + 4)$ and there are three of these. The area of the kitchen and dining room is $(x + 2)(x + 2)$ and there are two of these. The total area A is the sum of three algebraic expressions as follows.

Step 1: $A = 2x(x + 2) + 3(x + 2)(x + 4) + 2(x + 2)(x + 2)$
Step 2: $= (x + 2)[2x + 3(x + 4) + 2(x + 2)]$
Step 3: $= (x + 2)[7x + 16]$

In the second of the three steps, the common factor $(x + 2)$ was separated and factored out from all three terms. The third step is called the factored form for the area because the area is expressed as the product of two factors. It was derived by using the distributive property and then combining like terms.

A term such as $x + 2$ is called a **binomial** because it consists of two parts. The product $(x + 2)(x + 4)$ is called a product of two binomials. The multiplication of two binomials is a very important operation in algebra. It may be accomplished by a method known as the ***FOIL* method.**

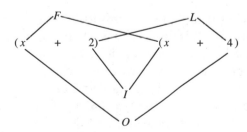

The *FOIL* method gets its name from the following: The **F**irst terms in the binomials are multiplied, the **O**utside terms are multiplied, the **I**nside terms are multiplied, and the **L**ast terms are multiplied.

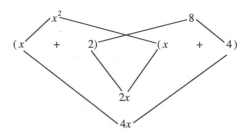

Combining the like terms $2x$ and $4x$ to get $6x$, we obtain the following.

$$(x+2)(x+4) = x^2 + 6x + 8$$

Using this technique, we find that

$$A = 2x(x+2) + 3(x+2)(x+4) + 2(x+2)(x+2)$$
$$= 2x^2 + 4x + 3(x^2 + 6x + 8) + 2(x^2 + 4x + 4)$$
$$= 2x^2 + 4x + 3x^2 + 18x + 24 + 2x^2 + 8x + 8$$
$$= 7x^2 + 30x + 32$$

We now have two expressions for the area: $A = (x + 2)(7x + 16)$ and $A = 7x^2 + 30x + 32$. We know that $7x^2 + 30x + 32 = (x + 2)(7x + 16)$. The expression $(x + 2)(7x + 16)$ is called the **factorization** of $7x^2 + 30x + 32$. That is, we have expressed $7x^2 + 30x + 32$ as a product of the two factors $x + 2$ and $7x + 16$.

Factoring is a very important part of basic algebra. To make certain that this discussion is clear, suppose x is 12 feet. Then the two upstairs bedrooms each have area $12(14) = 168$ square feet. The three downstairs bedrooms each have area $14(16) = 224$ square feet. The kitchen and dining room each have area $14(14) = 196$ square feet. The total is then $2(168) + 3(224) + 2(196) = 1400$. This total can also be found by using $7x^2 + 30x + 32 = (x + 2)(7x + 16)$ when $x = 12$.

$$7(12)^2 + 30(12) + 32 = 1400$$

$$(12 + 2)[7(12) + 16] = 14 \times 100 = 1400$$

Factoring

Example 1-17: A **trinomial** consists of three terms and can be factored by using the *FOIL* method for multiplying binomials *in reverse*. Suppose we wish to factor $x^2 + 3x + 2$. If the first terms are both x, we see that their product is x^2. That is $F = x(x) = x^2$. If the last terms are 1 and 2, their product is 2. That is $L = 1(2) = 2$. This suggests that the product $(x + 1)(x + 2)$ is possibly the factorization for $x^2 + 3x + 2$. The factorization is correct if the inside and outside products add to $3x$. The product of the inside terms is $I = x(1) = x$ and the product of the outside terms is $O = x(2) = 2x$ and their sum is $3x$. We see that $(x + 1)(x + 2)$ is the correct factorization for $x^2 + 3x + 2$.

Example 1-18: The factorization in **Example 1-17** is straightforward and relatively easy to follow. When the first term of the trinomial is x^2, it is usually easy to use the *FOIL* method in reverse. Suppose we wish to factor $4x^2 + 4x - 3$. The product of the first terms is $F = 4x^2$ and the product of the last terms is $L = -3$. In addition, $O + I$ must equal $4x$. Table 1-3 lists some possibilities for first terms and last terms that satisfy $F = 4x^2$ and $L = -3$. Note that six more possibilities would be obtained by replacing the terms in the a and c columns by their negatives.

Table 1-3 Listing Some Possibilities

$(a + b)(c + d)$								
a	b	c	d	F	O	I	L	$O + I$
$4x$	1	x	-3	$4x^2$	$-12x$	x	-3	$-11x$
$4x$	-1	x	3	$4x^2$	$12x$	$-x$	-3	$11x$
x	1	$4x$	-3	$4x^2$	$-3x$	$4x$	-3	x
x	-1	$4x$	3	$4x^2$	$3x$	$-4x$	-3	$-x$
$2x$	1	$2x$	-3	$4x^2$	$-6x$	$2x$	-3	$-4x$
$2x$	-1	$2x$	3	$4x^2$	$6x$	$-2x$	-3	$4x$

From Table 1-3, it is seen that the last row, shown in bold, is the only one that gives $O + I = 4x$. The correct factorization is then as follows.

$$4x^2 + 4x - 3 = (2x - 1)(2x + 3)$$

With practice, you can perform an equivalent process in the "mind's eye" that the table illustrates.

Example 1-19: Factor $17x^2 - 24x + 7$. We need the product of the first terms to equal $17x^2$; that is, $F = 17x^2$ and the product of the last terms to equal 7. We also need to have $O + I = -24x$. Table 1-4 lists some of the possibilities, with the correct one shown in bold.

Table 1-4   Listing Some Possibilities

	$(a + b)(c + d)$							
a	b	c	d	F	O	I	L	$O + I$
$17x$	1	x	7	$17x^2$	$119x$	x	7	$120x$
$17x$	7	x	1	$17x^2$	$17x$	$7x$	7	$24x$
x	1	$17x$	7	$17x^2$	$7x$	$17x$	7	$24x$
x	7	$17x$	1	$17x^2$	x	$119x$	7	$120x$
$17x$	-1	x	-7	$17x^2$	$-119x$	$-x$	7	$-120x$
$17x$	**-7**	**x**	**-1**	**$17x^2$**	**$-17x$**	**$-7x$**	**7**	**$-24x$**
x	**-1**	**$17x$**	**-7**	**$17x^2$**	**$-7x$**	**$-17x$**	**7**	**$-24x$**
x	-7	$17x$	-1	$17x^2$	$-x$	$-119x$	7	$-120x$

From Table 1-4, we see that either $(17x - 7)(x - 1)$ or $(x - 1)(17x - 7)$ is a correct factorization for $17x^2 - 24x + 7$.

Solving Equations

Example 1-20: This example illustrates one of the most frequently asked questions that teachers/professors encounter from their students. Suppose the grade in an algebra course is determined by four tests worth 100 points each and a final exam that counts 200 points. A student has scored 70, 55, 79, and 85 on the four tests. The student asks "What must I score on the final in order to obtain a C in the course?" Assuming 70% of the total points are needed to receive a C, the answer is found as follows.

Suppose we let x represent the number of points required on the final. The sum $70 + 55 + 79 + 85 + x = 289 + x$ must be at least 70% of the total possible. Because the total possible is 600, this leads to the following equation.

$$\frac{289 + x}{600} = 0.70$$

To solve this equation, we multiply both sides of the equation by 600 to obtain

$$289 + x - 600(0.70)$$
$$289 + x = 420$$

By subtracting 289 from both sides of $289 + x = 420$, we find that $x = 131$. That is the student must score at least 131 points on the final exam to receive at least a C in the course. If the final is expressed as a percent, the student must score at least $131/200(100) = 66\%$ on the final to obtain a C in the course.

Short Cuts

The recent development of computer algebra software has added a new dimension to the teaching of algebra at all levels. One of the widely used computer algebra packages is MAPLE (Waterloo Maple Inc., 57 Erb Street West, Waterloo, Ontario, Canada N2L 6C2). This program is available at many universities and a version for use in high school and community college mathematics is currently being investigated. I have used MAPLE to supplement the teaching of my intermediate algebra course and it has been well received by my students. (The results of a study concerning this usage is reported in a paper entitled "The Use of Computer Algebra in Teaching Intermediate and College Algebra" by Larry J. Stephens and John Konvalina in the *International Journal of Mathematical Education in*

Science and Technology, vol. 30, no. 4, pp. 483–488.) Output from this software will be included at various points in this book. The MAPLE solution to Example 1-20 is as follows: >solve((289 + *x*)/600 = .70); is the command in MAPLE to solve the equation. 131 is the answer provided by MAPLE, the same as obtained in Example 1-20.

Don't Forget

The real numbers are best visualized as corresponding to points lying along a line extending from very small negative numbers on the left side to very large positive numbers on the right.

Operations on Fractions

1. To add or subtract fractions, first change all fractions to equivalent fractions having a common denominator. Then add or subtract the numerators and put the result over the common denominator.

2. To multiply fractions, multiply the numerators and put the result over the product of the denominators.

3. To divide fractions, invert the fraction you are dividing by and then multiply.

Fundamental Operations Involving Signed Numbers

1. The sum of two positive numbers is positive and the sum of two negative numbers is negative.

2. The sum of a positive and a negative number is positive if the larger number (in absolute value) is positive, and the sum is negative if the larger number (in absolute value) is negative. The sum of a positive and a negative number having the same absolute value is zero.

3. To subtract two signed numbers, add the opposite of the second signed number.

4. The product of a positive number and a negative number is negative. The product of two positive numbers or the product of two negative numbers is positive.

5. The quotient of a positive and a negative number is negative and the quotient of two negative numbers or two positive numbers is positive.

A number expressed as a product of a number between 1 and 10 (not including 10) or −1 and −10 and some power of 10 is said to be in scientific notation.

When you multiply two numbers in exponential notation having the same base, you add exponents and when you divide two numbers in exponential notation having the same base, you subtract exponents.

Each positive integer is either a prime number or it is expressible as a product of primes.

A variable is a letter that is used to represent a number. Examples of the use of variables in area and volume formulas are as follows:

The area A of a rectangle having length l and width w is $A = lw$.

The area A of a square having a side equal to s is $A = s^2$.

The area A of a circle having radius r is $A = \pi r^2$.

The volume V of a rectangular solid having length l, width w, and height h, is $V = lwh$.

The volume V of a cube having a side equal to s is $V = s^3$.

The volume V of a right circular cylinder having radius r and height h is $V = \pi r^2 h$.

Different algebraic expressions may be combined by the operations of addition, subtraction, multiplication, and division.

A solution to an equation is a number that makes the equation true when the number is substituted for the variable in that equation.

Test Yourself

Questions

1. Give the decimal equivalents for the Wednesday 52-week high, 52-week low, close, and change for Pfizer shown in Table 1-1.

2. Use the Wednesday and Thursday closing values in Table 1-1 to obtain the change value given on Thursday for a share of IBM.

3. Suppose the Exxon closing value in Table 1-1 was split into halves on Wednesday. What would each new share be worth?

4. Find the percent change for GE on Thursday.

5. Evaluate $\dfrac{7}{16} + \dfrac{4}{32} - \dfrac{36}{64}$.

6. Evaluate $\dfrac{7}{36}\left(\dfrac{72}{14}\right)$.

7. Evaluate $\dfrac{17}{55} \div \dfrac{34}{165}$.

8. Suppose the daily changes for AT&T stock were $+1, -3, +4, +1.5,$ and $-3\dfrac{7}{16}$ during a given week. What was the change for the week?

9. Suppose Coca-Cola stock is down 3 points for each day in a given week. What was the change for the week?

10. Evaluate -500 divided by -125.

11. A total of 2,600,000,000 prescriptions were filled in the United States last year. Express this number in scientific notation.

12. Referring to **Problem 11,** how many prescriptions were filled per day last year. Assume an equal number were filled per day and express your answer in scientific notation.

13. Evaluate $(3^3)(2^4)$.

14. Give the prime factorization for 30,240.

15. Give the prime factorization for 15,015.

16. Use the FOIL method on the two separate parts of the following expression to find a simplified expression for it: $(x+3)(2x-5) - (2x+1)(x+3)$

17. Factor $x^2 + x - 42$.

18. Factor $6x^2 - 7x - 5$.

19. Factor $3x^2 + 10x - 8$.

20. Suppose your grade is determined from four 100-point exams. You have scored 85, 70, and 60 on the first three and it is required that your average be at least 80 on the four tests in

order to receive a B in the course. Is it possible to receive a B in the course?

Answers

1. 134.875, 84.125, 132.5, and −1.375

2. $-\dfrac{1}{8}$

3. $33\dfrac{119}{128}$

4. −0.62%

5. 0

6. 1

7. 1.5

8. $\dfrac{1}{16}$

9. −15

10. 4

11. 2.6×10^{9}

12. 7.1×10^{6}

13. 432

14. $2^{5}(3^{3})(5)(7)$

15. $3(5)(7)(11)(13)$

16. $-6x - 18$

17. $(x - 6)(x + 7)$

18. $(3x - 5)(2x + 1)$

19. $(3x - 2)(x + 4)$

20. No, you would need 105 points to average 80% on the four tests.

CHAPTER 2

Properties of Numbers

Do I Need to Read This Chapter?

➜ What is the distributive property of numbers?

➜ What is the additive inverse of a number?

➜ What is the multiplicative inverse of a number?

➜ What is the additive identity?

➜ What is the multiplicative identity?

➜ What is the commutative law of addition?

➜ What is the associative law of addition?

➜ What is the commutative law of multiplication?

➜ What is the associative law of multiplication?

➜ What is the multiplication property of zero?

➜ What is the proper order of operations?

Distributive Property

Example 2-1: A real estate management company owns an apartment complex. The complex contains three two-bedroom apartments and three three-bedroom apartments. The two-bedroom apartments rent for $750 per month and the three-bedroom ones rent for $1000 per month. Compute the monthly rent from the six apartments in two different ways.

One method would be to add the cost of a two-bedroom apartment and a three-bedroom apartment to get $1750 and then to multiply $1750 by 3 to get $5250. This could be represented by 3($750 + $1000). An equivalent method is to multiply the rent for a two-bedroom apartment by 3 to get $2250, multiply the rent for a three-bedroom apartment by 3 to get $3000, and then add these two to obtain $5250. This could be represented by 3($750) + 3($1000). This discussion is summarized by the following equation.

$$3(\$750 + \$1000) = 3(\$750) + 3(\$1000)$$

This computation illustrates the **distributive property** of numbers. This property of numbers is widely used in algebra. This property is generalized as follows: For any numbers *a, b,* and *c,*

$$a(b + c) = ab + ac$$

Danger!

A common mistake that is often made in applying the distributive law is the following, especially if a calculator is being used. To evaluate 3($750 + $1000), a 3 will be entered into the calculator, followed with multiplication by $750, followed by the addition of $1000. This will result in 3(750) + 1000 = 3250, which, of course, is not correct. The parentheses indicate that the addition should be completed first and then be followed by the multiplication. The following order of operations may also result in an incorrect answer. The addition of $750 and $1000 is immediately followed by a multiplication by 3. If the addition is not completed before the multiplication by 3, the answer 750 + 1000(3) = 3750 is obtained. In this case $750 was added to $1000 times 3 which yields $3750. To ensure that the addition is accomplished before the multiplication by 3, hit the equal button on the calculator after the addition and before multiplying by 3. Also, if the calculator has parentheses, these can be used to ensure the proper order of operations. Table 2-1 gives some examples of the application of the distributive law.

Table 2-1

EXPRESSION	EQUIVALENT EXPRESSION
$10(5 + 5)$	$50 + 50$
$X(A + B)$	$XA + XB$
$5(4 - 2)$	$20 - 10$
$W(2X + 3Y)$	$2WX + 3WY$
$4A(2B - 3C)$	$8AB - 12AC$

Additive Inverse

Example 2-2: Suppose you deposit $150 into your checking account and then use your debit card to purchase a CD player for $150. What net result will these two transactions produce?

The deposit can be represented by $150 and the debit card purchase can be represented by –$150. The net result is $150 – $150 = 0. The algebra term that describes this is additive inverse. We can say that the debit card purchase is the additive inverse of the deposit, or that the deposit is the additive inverse of the debit card purchase. In general, for any number a, there always exist another number $-a$, called the **additive inverse** of a, for which the following is true.

$$a + (-a) = 0$$

Table 2-2 gives some quantities and their additive inverses.

Table 2-2

QUANTITY	ADDITIVE INVERSE
20	−20
−50	50
X	$-X$
$X + Y$	$-X - Y$
$1.50	−$1.50

Multiplicative Inverse

Example 2-3: Suppose five powerball winners are to share $1,000,000. The share per winner can be determined by dividing the $1,000,000 by 5 to obtain $200,000 each. Is there a number we could multiply times $1,000,000 to obtain the share per winner? Yes, there is such a number. It is called multiplicative inverse of 5 and is equal to 1 divided by 5, or $\frac{1}{5}$. This $\frac{1}{5}$ written as a decimal is 0.2. Note that 0.2 times $1,000,000 also gives $200,000.

Quick Tip

One practical use of the multiplicative inverse involves its use to express quantities on a per unit basis.

Example 2-4: For example suppose 10 pounds of bananas cost 90 cents. The cost per pound can be found by multiplying 90 cents by the multiplicative inverse of 10 or $\frac{1}{10} = 0.1$. The cost per pound is 90 (0.1) or 9 cents per pound.

Example 2-5: Suppose 2 ounces ($\frac{1}{8}$ pound) of black pepper cost 75 cents. The cost per pound can be found by multiplying 75 cents by the multiplicative inverse of $\frac{1}{8}$. (If $\frac{1}{8}$ is divided into 1 the result is 8). The multiplicative inverse of $\frac{1}{8}$ is 8. The cost per pound is $75(8) = \$6.00$. Table 2-3 gives various quantities and their multiplicative inverses.

The **multiplicative inverse** (for any number other than zero) is found by dividing the number into 1. If a represents a nonzero number, then the multiplicative inverse for a is $\frac{1}{a}$.

Table 2-3

QUANTITY	MULTIPLICATIVE INVERSE
10	0.10
$0.02 = \frac{2}{100}$	$\frac{100}{2} = 50$
$-0.2 = -\frac{2}{10}$	$-\frac{10}{2} = -5$
X, when $X \neq 0$	$1/X$
$1/Y$, when $Y \neq 0$	Y

Additive Identity

The numbers 0 and 1 play such important roles in algebra that they are given special names. The number 0 is called the **additive identity.** It has the property that when it is added to any other number, the result is the original number. The original number keeps its "identity." For any number a, we have the following equation.

$$a + 0 = 0 + a = a$$

Multiplicative Identity

The number 1 is called the **multiplicative identity.** It has the property that when it is multiplied times any other number, the result is the original number. The original number keeps its "identity." For any number *a,* we have the following equation.

$$a(1) = 1(a) = a$$

Table 2-4 gives several examples of algebraic operations involving the additive identity and the multiplicative identity.

Table 2-4

ALGEBRAIC OPERATION	RESULT
$12 + 0$	12
$M(1)$	M
$X + 0$	X
$Y(1) + 0$	Y
$Y(1 + 0)$	Y

Commutative Law of Addition

Example 2-6: Larry and Lana had a lucky day at the casino. Lana played video poker and won $2000 and Larry played the slot machine called Flaming Sevens. He won $3000.

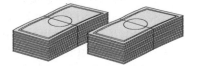 

$$\$2000 + \$3000 = \$3000 + \$2000$$

Whether Lana cashes in first and Larry second or Larry cashes in first and Lana second, the total is the same. We refer to this fact in algebra as the commutative law of addition. In general, if *a* and *b* are any numbers, then the **commutative law of addition** states that

$$a + b = b + a$$

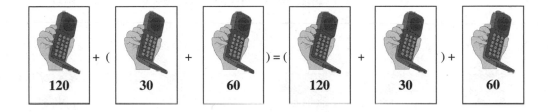

Associative Law of Addition

120 minutes + (30 minutes + 60 minutes)

= (120 minutes + 30 minutes) + 60 minutes

Example 2-7: Three family members each have a cell phone. The monthly uses for the three members are 120, 30, and 60 minutes. In determining the total time, you can add 30 and 60 to get 90 and then add 90 to 120 to obtain 210 minutes or you can add 120 and 30 to get 150 and then add 150 and 60 to get 210 minutes. The result is the same for each. This illustrates the associative law of addition.

In general, if *a, b,* and *c* are any numbers, then the **associative law of addition** is:

$$a + (b + c) = (a + b) + c$$

Commutative Law of Multiplication

Example 2-8: A nursery sells roses for \$1.50 each. To determine the cost of half a dozen roses, you can multiply 6 times the cost of one rose to obtain 6(\$1.50) = \$9.00 or you can multiply the cost of one times 6 to obtain \$1.50(6) = \$9.00. In general, if *a* and *b* are any numbers, then the **commutative law of multiplication** states that

$$a(b) = b(a)$$

Associative Law of Multiplication

Example 2-9: If one rose costs \$1.50, how much would three dozen roses cost? We could first figure the number of roses in three dozen to obtain $3 \cdot 12 = 36$. Then multiply 36 times the cost per rose to obtain $(3 \cdot 12)\$1.50 = \$54$. Or, we could figure the cost of a dozen as 12(\$1.50) = \$18. Then follow this by multiplying by the number of dozens to obtain $3(12 \cdot \$1.50) = \$54$. If *a, b,* and *c* are real numbers, then the **associative law of multiplication** states that

$$a(b \cdot c) = (a \cdot b)c$$

Multiplication Property of Zero

The **multiplication property of zero** states that when zero is multiplied by any real number *a*, the result is zero, that is,

$$a(0) = 0(a) = 0$$

Pattern

Expressions are encountered in algebra that require the utilization of many of the properties discussed in this chapter. The proper *order of operations* in such expressions is as follows.

1. Evaluate inside grouping symbols—parentheses ( ), brackets [ ], braces { }, or absolute values | |—first. Also evaluate expressions above or below a fraction bar or under a radical.

2. Evaluate powers, roots, and absolute values.

3. Evaluate multiplications and divisions as they occur from left to right.

4. Evaluate additions and subtractions as they occur from left to right.

Example 2-10: Suppose we wish to evaluate the expression: $2[(3 + 4)(4) - 2(3)(5) - \{10 - 6\}]$ The proper steps to follow in evaluating the expression are as follows:

$2[(3 + 4)(4) - 2(3)(5) - \{10 - 6\}]$

$= 2[(7)(4) - 2(3)(5) - \{4\}]$ (Evaluate inside parenthesis and braces.)

$= 2[28 - 30 - 4]$ (Evaluate multiplications and divisions from left to right.)

$= 2[-6]$ (Evaluate additions and subtractions from left to right.)

$= -12$

After discussing powers, roots and absolute values in the following chapters, we want to remember the order of operations as summarized above.

Short Cuts

The ultimate in speed for evaluating expressions in algebra is provided by computer software. Consider the solution to Example 2-10 produced by the software package MAPLE. This software is available for student use at many colleges and universities. After accessing the software, the prompt provided is the character >. Study the syntax for the command and identify the characters used for multiply, divide, etc.

```
> 2*[(3 + 4)*(4) - 2*3*5 - (10 - 6)];        [-12]
```

Don't Forget

The following laws need to be practiced by working problems from your algebra text until you have them memorized. They need to become a part of your algebra vocabulary.

The distributive law: $a(b + c) = ab + ac$

The additive inverse: $a + (-a) = 0$

The multiplicative inverse: $a(1/a) = a/a = 1, a \neq 0$

The additive identity: $a + 0 = 0 + a = a$

The multiplicative identity: $a(1) = 1(a) = a$

The commutative law for addition: $a + b = b + a$

The associative law for addition: $a + (b + c) = (a + b) + c$

The commutative law for multiplication: $a(b) = b(a)$

The associative law for multiplication: $a(b \cdot c) = (a \cdot b)c$

The multiplication property of zero: $a(0) = 0(a) = 0$

Test Yourself

Questions

1. Use the distributive property of numbers to find an equivalent expression for the following:
   (a) $2(3 + 5)$
   (b) $x(y + z)$
   (c) $3a(2b + c)$

2. Give the additive inverse for the following:
   (a) -575
   (b) $-2x + 4W$
   (c) $\$75$

3. Give the multiplicative inverse for the following:
   (a) 0.5
   (b) -5
   (c) $x + 2y \neq 0$

4. If a box of 10 floppy disks cost $5, what is the cost per floppy? Use the multiplicative inverse to find the answer.

5. If raw sunflower seeds are 10 cents per ounce, what would one-half pound cost? Use the multiplicative inverse to find the answer.

6. Use the commutative law of addition to give an equivalent expression for the following:

   (a) $4 + 8$

   (b) $2x - 4s$

   (c) $a + b$

7. Use the associative law of addition to give an equivalent expression for the following:

   (a) $4 + (2 + 5)$

   (b) $(4 + 2) + 5$

8. Use the commutative law of multiplication to give an equivalent expression for the following:

   (a) $25(78)$

   (b) $78(25)$

9. Use the associative law of multiplication to find an equivalent expression for the following:

   (a) $33(13 \cdot 12)$

   (b) $(2x \cdot 4x)z$

10. Evaluate the expression $[2(3) - 5/5 + 6(3 + 4(7))]$.

Answers

1. (a) $6 + 10$ (b) $xy + xz$ (c) $6ab + 3ac$

2. (a) 575 (b) $2x - 4W$ (c) $-\$75$

3. (a) 2 (b) -0.2 (c) $\dfrac{1}{x + 2y}$

4. 50 cents

5. 80 cents

6. (a) $8 + 4$ (b) $-4s + 2x$ (c) $b + a$
7. (a) $(4 + 2) + 5$ (b) $4 + (2 + 5)$
8. (a) $78(25)$ (b) $25(78)$
9. (a) $(33 \cdot 13)12$ (b) $2x(4x \cdot z)$
10. 191

Functions

Do I Need to Read This Chapter?

→ What is a function and what are the domain and range of a function?

→ What is a one-to-one function?

→ What is a rectangular coordinate system?

→ What are Minitab and Excel and how can they enhance the study of algebra?

→ What is a linear function and what is the slope and y-intercept associated with a linear function?

→ What is an inverse function?

→ What are the vertical and horizontal line tests?

Function, Domain, and Range

Example 3-1: The National Football League is composed of the American Football Conference (AFC) and the National Football Conference (NFC). There are three divisions in each conference and five teams in each division. Table 3-1 gives the standings for the West division of the AFC.

Table 3-1   Standings in the West Division of the *AFC*

TEAM	WON	LOST	PCT
Denver	11	0	1.000
Oakland	7	4	.636
San Diego	5	6	.455
Seattle	5	6	.455
Kansas City	4	7	.364

A **function** is a rule that assigns to each element in a set, called the **domain,** exactly one element in a second set, called the **range.** In Table 3-1, if the domain is the set of teams in the West division of the AFC and the range is the set of percentages in the pct column, then the rule that connects the two sets can be described as follows. For a given team, the pct is formed by dividing the number of games won by the total number of games played. (Note that the pct values are not actually percentages. They would need to be multiplied by 100 to actually be percentages.) The elements in the domain of the function are often represented by the letter x and the elements of the range are represented by the letter y. The functional relationship may be expressed in equation form as follows.

$$y = f(x)$$

The function notation does not mean f times x. The notation, $f(x)$, is read "f of x." It tells us how y is determined when x is specified. If

x = Oakland, then f(Oakland) = 7/11 = 0.636. If x = Denver, then f(Denver) = 11/11 = 1.000. The other three pct values are found in the same manner.

Quick Tip

Examples of the concept *function* are all around us in the real world. The NYSE, NASDAQ, and AMEX stock listings as well the mutual funds listings may be regarded as functions. For the stock listings, the companies comprise the domain. There are several different ranges such as the closing value, change, and the price to earnings ratio (P/E). The domain for the mutual funds is the different funds. Ranges are measures such as NAV (net asset value), year-to-date total return percent, and change.

One-to-One Function

If there is exactly one element in the range for each element in the domain and if there is exactly one element in the domain for each element in the range, then we say that the function is a **one-to-one function.** The function described in Table 3-1 is not a one-to-one function because the element .455 in the range corresponds to both San Diego and Seattle. That is, these two teams have identical won and lost records and hence the same pct values. If all five teams had different won/lost records, then the function would be one-to-one.

Example 3-2: If the Centigrade temperatures are the elements of the domain and the Fahrenheit temperatures are the elements of the range, then the values in Table 3-2 represent a functional relationship, because there is exactly one value in the range for each domain value. Furthermore, the function is a one-to-one function because it is also true that for each range value (Fahrenheit temperature) there is only one domain value (Centigrade temperature).

Table 3-2 Fahrenheit and Centigrade Temperatures

DOMAIN X = CENTIGRADE	RANGE Y = FAHRENHEIT
0*	32*
10	50
20	68
30	86
40	104
50	122
60	140
70	158
80	176
90	194
100†	212†

*Freezing point of water.
†Boiling point of water.

Rectangular Coordinate System

Example 3-3: Functions are often represented by plotting pairs of values such as those shown in Table 3-2 in a **rectangular coordinate system.** A plot of these points is shown in Fig. 3-1. This plot was produced by the software package Minitab. Note that each plotted point represents a given pair of values from Table 3-2.

Figure 3-2 gives many of the terms associated with a rectangular coordinate system. The coordinate system is composed of two perpendicular lines called the **x-axis** and the **y-axis.** The x-axis and the y-axis meet at a point called the **origin.** The four **quadrants** are numbered in a counterclockwise manner as shown in the figure. Figure 3-1 is actually a portion of the first quadrant only. In this figure, the x-axis, also called Centigrade, is scaled from 0 to 100 and the y-axis, is called Fahrenheit and scaled from 20 to 220. As indicated in Fig. 3-2, in quadrant I, the x-value is positive and the y-value is positive.

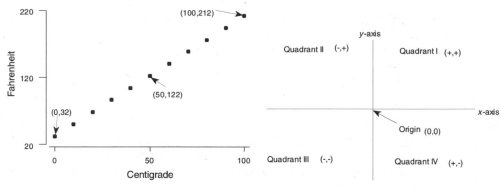

Fig. 3-1   **Fig. 3-2**

In quadrant II, the *x*-value is negative and the *y*-value is positive, in quadrant III, the *x*-value is negative and the *y*-value is negative, and in quadrant IV, the *x*-value is positive and the *y*-value is negative.

Example 3-4: The **bar graph** is another popular device used to represent a function. Fig. 3-3 is a bar graph produced by Minitab, giving the baseball teams with the most MVPs (most valuable player) prior to the 1998 season.

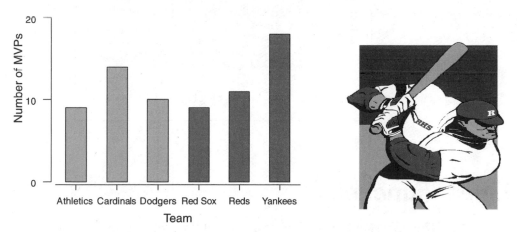

Fig. 3-3

The domain of this bar graph consists of the six teams and the range consists of the numbers 9, 10, 11, 14, and 18.

Thus far we have considered three different methods of illustrating a function: a table, a rectangular coordinate system, and a bar chart. There are many other methods. The functions considered have all consisted of a finite number of pairs. Often, we are interested in functions consisting of infinitely many pairs. A rectangular coordinate system is particularly useful in representing such functions. Figure 3-1 shows the relationship between 11 pairs of Centigrade and Fahrenheit temperatures. If these points are connected by use of a ruler, a graph of the infinite number of pairs of possible corresponding points is obtained. The **Microsoft Excel spreadsheet** graph obtained when the points in Fig. 3-1 are connected by use of a ruler is shown in Fig. 3-4. This graph shows the infinite number of pairs of points that exist between the Centigrade and the Fahrenheit temperature scales.

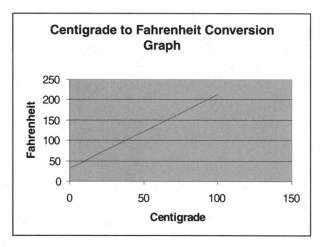

Fig. 3-4

Linear Functions

The points in Fig. 3-4 fall on a straight line. Such a function is said to be linear. A **linear function** can always be expressed as follows, where m is the **slope of the line** and b is the **y-intercept.**

$$f(x) = mx + b$$

Short Cuts

Example 3-5: One technique for determining *m* and *b* is now given. The *y*-intercept is the point where the line intercepts the *y*-axis. The *y*-intercept is the point (0, *b*). This point can easily be determined from Table 3-2. Note from Table 3-2 that when $x = 0$, $y = 32$, implying that $b = 32$. The slope is determined by choosing any two points on the line and dividing the difference in the *y*-values by the difference in the *x*-values. Suppose we choose the freezing and boiling points of water: (0, 32) and (100, 212). The slope *m* is found as follows:

$$m = \frac{212 - 32}{100 - 0} = 1.8$$

The equation of the linear function connecting Fahrenheit (*y*) to Centigrade (*x*) is:

$$y = f(x) = 1.8x + 32$$

Two widely used software packages have been introduced so far in this chapter. Both of these software packages are enhancements to the understanding of algebra. A third software package that is even more useful in helping to enhance the learning of algebra is the package called *MAPLE*. Consider the example of a vacationing family traveling at a constant velocity of 70 miles per hour along Interstate 80. In 1 hour the family travels 70 miles. In 2 hours, the family travels 140 miles. In 2-1/2 hours the family travels $140 + 35 = 175$ miles, and so forth. In general, the family travels 70*x* miles in *x* hours. That is, the functional relationship between time *x* and distance *y* is $y = f(x) = 70x$. A MAPLE plot for this function is shown in Fig. 3-5.

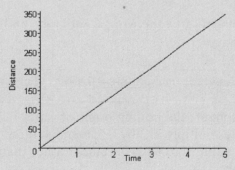

Fig. 3-5

Inverse Function

Example 3-6: When a function is a one-to-one function, there exists an **inverse function.** This concept can be illustrated with the temperature conversion example discussed earlier. Because there is exactly one Fahrenheit temperature corresponding to each Centigrade temperature and exactly one Centigrade temperature corresponding to each Fahrenheit temperature, the function $y = f(x) = 1.8x + 32$ or Fahrenheit = 1.8(Centigrade) + 32 can be solved for x as follows. First subtract 32 from both sides.

$$y - 32 = 1.8x$$

Next, divide both sides of this new equation by 1.8 to obtain the following.

$$\frac{y - 32}{1.8} = x$$

Simplifying the left-hand side we get

$$0.556y - 17.778 = x$$

Finally, the expression for x can be expressed as

$$x = 0.556y - 17.778$$

Replacing x by the word Centigrade and y by the word Fahrenheit, we have

Centigrade temperature = 0.556   Fahrenheit temperature – 17.778

To convert the boiling point of water in Fahrenheit, 212, to Centigrade multiply 212 by 0.556 and subtract 17.778 to obtain 100.094. The value 100.094 is obtained rather than 100 because of round-off error.

Short Cuts

Two visual tests of graphs that are often used to determine functions and their possible one-to-one relationship are the *horizontal line test* and the *vertical line test*. The vertical line test determines whether it is a function and the horizontal line test determines whether it is a one-to-one function. The vertical line test states that a graph represents a function if every vertical line intersects the graph in only one point. The horizontal line test states that if every horizontal line intersects the graph of the given function in no more than one point, then it is a one-to-one function and its inverse will also be a function.

Example 3-7: By considering Fig. 3-5, we note that any vertical line you could draw intersects the graph in exactly one point and that any horizontal line you could draw intersects the graph in exactly one point. The equation of the line in Fig. 3-5 is $y = 70x$, where x is time and y is distance. Solving for time in terms of distance, we obtain the equation

$$x = \frac{y}{70}$$

Or expressed in another way, we have

$$\text{Time} = \frac{\text{distance}}{70}$$

This is the inverse function for the function: distance = 70 time.

Pattern

The following three steps are recommended to find the equation of the inverse function for a one-to-one function.
Step 1: Exchange x and y in the given equation.
Step 2: Solve for y.

Step 3: Replace y by $f^{-1}(x)$. This notation is used to represent the inverse function that corresponds to the function $f(x)$. This notation represents the set of ordered pairs of the form (y,x), where (x,y) belongs to f. It does not represent 1 over $f(x)$.

Example 3-8: The equation for the function relating Centigrade temperature x to Fahrenheit y is $y = 1.8x + 32$.

Step 1: Exchange x and y in the given equation.

$$x = 1.8y + 32$$

Step 2: Solve for y.

$x - 32 = 1.8y$ (Add −32 to both sides of the equation.)

$\dfrac{x - 32}{1.8} = y$ (Divide both sides by 1.8.)

Step 3: Replace y by $f^{-1}(x)$.

$f^{-1}(x) = \dfrac{x - 32}{1.8}$, where x now represents Fahrenheit temperatures.

To convert the boiling point of water in Fahrenheit, 212, to Centigrade, we proceed as follows:

$$f^{-1}(212) = \frac{212 - 32}{1.8} = 100$$

Note that the function is represented by $f(x)$ and the inverse function is represented by $f^{-1}(x)$. In this example, $f(x) = 1.8x + 32$ is used to convert Centigrade to Fahrenheit and $f^{-1}(x) = \dfrac{x - 32}{1.8}$ is used to convert Fahrenheit to Centigrade.

Example 3-9: Use the three-step procedure to solve Example 3-7. The equation relating distance to time is $y = 70x$.

Step 1:   Exchange *x* and *y* in the given equation.

$$x = 70y$$

Step 2:   Solve for *y*.

$$\frac{x}{70} = y$$

Step 3:   Replace *y* with $f^{-1}(x)$.

$$f^{-1}(x) = \frac{x}{70}, \text{ where } x \text{ represents distance traveled}$$

and $f^{-1}(x)$ gives the time required.

Danger!

Beware! Not every tabular presentation of data represents a function

Example 3-10: Table 3-3 indicates that in 1995, four states executed 1 prisoner, four states executed 2 prisoners, two states executed 3 prisoners, two states executed 5 prisoners, one state executed 6 prisoners, and one state executed 19 prisoners. If the

Table 3-3   Prisoners Executed in 1995

DOMAIN X = NUMBER EXECUTED	RANGE Y = STATES
1	AZ, DE, LA, SC
2	AL, AR, GA, NC
3	FL, OK
5	IL, VA
6	MO
19	TX

Source: U.S. Bureau of Justice Statistics.

number executed is the domain and the state is the range, then this presentation does not represent a function because for each element in the domain there is not exactly one element in the range.

Don't Forget

A *function* is a rule that assigns to each element of one set called the *domain*, exactly one element of a second set called the *range*.

A *one-to-one function* is a function in which each value of *x* is paired with exactly one value of *y* and each value of *y* is paired with exactly one value of *x*.

A *rectangular coordinate system* consists of two perpendicular lines called the *x-axis* and the *y-axis*, a center point called the *origin*, and four *quadrants*.

A *bar graph* can be used to plot a function in which the domain consists of categories and the range consists of the frequency of occurrence for the categories.

When a function is a one-to-one function, there exists an *inverse function.* The domain of the inverse function is the same as the range of the original function and the range of the inverse function is the same as the domain of the original function.

If every vertical line you could draw intersects a graph in only one point, then the graph represents a function. This is referred to as the *vertical line test.*

If each horizontal line that can be drawn intersects the graph of a given function in no more than one point, then it is a one-to-one function and its inverse will also be a function. This result is referred to as the *horizontal line test.*

Test Yourself

Questions

1. In Table 3-4, if the domain is the players and the range is the number of home runs, would that portion of the table consisting of players and home runs represent a function? If it is a function, is it a one-to-one function?

Table 3-4   Five Top Home Run Hitters of All Time

PLAYER	HOME RUNS	POSITIONS PLAYED
Hank Aaron	755	Outfield
Babe Ruth	714	Outfield, pitcher
Willie Mays	660	Outfield
Frank Robinson	586	Outfield, designated hitter, first base
Harmon Killebrew	573	First base, third base, outfield, designated hitter

2. In Table 3-4, if the domain is the players and the range is the positions played, would that portion of the table represent a function? If it is a function, is it a one-to-one function?

3. Consider the function composed of the ordered pairs {(1, 2), (−2, 3), (−1, −3), (3, −4)}.
   (a) Give the domain for this function.
   (b) Give the range for this function.
   (c) If the ordered pairs are plotted, what quadrant of the rectangular coordinate system would each fall in?

4. Figure 3-6 shows an Excel-generated pie chart. What is the domain and range for this function,?

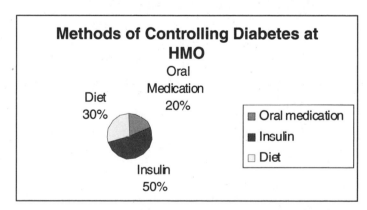

Fig. 3-6

5. For the linear function $f(x) = 2x + 4$, where does the graph of this function intercept the y-axis? What is the slope of the line?

6. If $f(x) = 3x - 3$, find $f(3)$.

7. If $f(x) = x + 4$, find $f^{-1}(x)$.

8. In Problem 7, find $f(6)$ and $f^{-1}(10)$.

9. If $f(x) = 10x - 5$, find $f^{-1}(x)$.

10. In Problem 9, find $f(2)$ and $f^{-1}(15)$.

Answers

1. It is a one-to-one function.

2. Not a function

3. (a) Domain consists of $-2, -1, 1$, and 3.   (b) Range consists of $-4, -3, 2$, and 3.   (c) I, II, III, and IV, respectively.

4. Domain consists of diet, oral medication, and insulin. Range consists of $20\%, 30\%$, and 50%.

5. $(0, 4)$ slope $= 2$.

6. 6

7. $f^{-1}(x) = x - 4$.

8. $f(6) = 10$ and $f^{-1}(10) = 6$.

9. $f^{-1}(x) = (x + 5)/10$.

10. $f(2) = 15$ and $f^{-1}(15) = 2$.

CHAPTER 4

Linear Functions and Linear Equations

Do I Need to Read This Chapter?

➡ What is the slope of a straight line?

➡ What are the x- and y-intercepts of a straight line?

➡ How can I use the slope to determine if two lines are parallel or perpendicular?

➡ What are the multiplication and addition principles associated with solving linear equations?

Linear Functions

Example 4-1: In Chapter 3, we saw that a linear function has an equation of the form $y = mx + b$ and has a graph that is a straight line. Telephone bills are made up of two parts, a fixed cost as well as a variable part. Suppose the local communications cost is equal to $25.00 per month and that the long-distance component cost consists of a charge of 15 cents per minute for long-distance calls within the continental United States. If x represents the number of long-distance minutes per month, and y represents the total amount of the monthly bill, then y may be expressed as $y = 25 + 0.15x$. A plot of the function

$$y = 25 + 0.15x$$

$$\text{Cost} = 25 + 0.15 \text{ (long-distance minutes)}$$

is shown in Fig. 4-1.

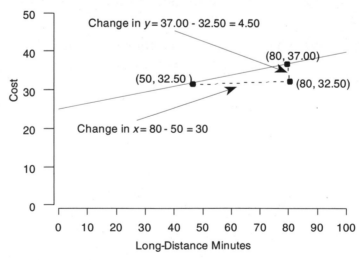

Cost as a Function of Long-Distance Minutes

Fig. 4-1

Slope of a Straight Line

Example 4-2: Two points are shown on the graph relating cost to long-distance minutes in Fig. 4-1. For $x = 50$ minutes, the cost is $y = 25 + 0.15(50) = \$32.50$ and for $x = 80$ minutes, the cost is $y = 25 + 0.15(80) = \$37.00$. The **change in x** and the **change in y** for these two points are shown in the figure. The ratio of the change in y to the change in x is called the **slope** of the line. From Fig. 4-1 we have the following:

Change in $x = 80 - 50 = 30$ and Change in $y - 37.00 - 32.50 = 4.50$

The slope is found as follows:

$$m = \frac{\text{change in } y}{\text{change in } x} = \frac{4.50}{30} = 0.15$$

Note that the slope is the same as the long-distance cost per minute. The slope tells us that for a 1-minute increase in long-distance time, the cost increases 0.15 dollars or 15 cents.

If (x_1, y_1) and (x_2, y_2) represent any two points on a straight line then the slope may be found by using the following equation:

$$m = \frac{\text{change in } y}{\text{change in } x} = \frac{y_2 - y_1}{x_2 - x_1} = \frac{y_1 - y_2}{x_1 - x_2}$$

Danger!

In calculating the slope, it does not matter which point you label (x_1, y_1) and which you label (x_2, y_2). For the slope calculation related to Fig. 4-1, if we compute change in $x = 50 - 80 = -30$ and change in $y = 32.50 - 37.00 = -4.50$, then $m = -4.50/(-30) = 0.15$ and we obtain the same slope as computed previously. However, *do not use different orders of the points in computing the change in x and for computing the change in y. This will give you the slope with an incorrect sign.*

x- and *y*-Intercepts of a Straight Line

  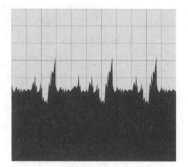

Example 4-3: Suppose that you inherit $50,000 and are trying to decide what part to invest in the stock market and what part to invest in mutual funds. Suppose x represents the amount to be invested in the stock market and y represents the amount to be invested in mutual funds. Then x and y must satisfy the following equality $x + y = 50,000$. An equivalent way of expressing this relation is $y = 50,000 - x$. A Minitab plot of the possible allocations to stocks and mutual funds is shown in Fig. 4.2. The point where the

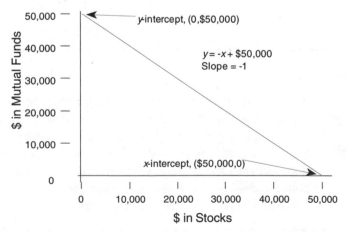

Fig. 4-2

line cuts the x-axis is called the **x-intercept** and the point where the line cuts the y-axis is called the **y-intercept.** The x-intercept is found by setting y equal to zero and solving the equation $y = 50,000 - x$ for x. This is seen to equal 50,000. This would represent the amount put into stocks if nothing were put into mutual funds. The y-intercept is found by setting $x = 0$ and solving the equation for y. This is also seen to be equal to 50,000. Thus, if nothing is put into stocks, then 50,000 is put into mutual funds. Comparing the equation $y = -x + 50,000$ to the general linear equation $y = mx + b$, we see that the slope is equal to -1. This tells us that for each increase of \$1 put into stocks there is \$1 less available to put into mutual funds.

Remember, to find the x-intercept, set y equal to zero in the linear equation and solve for x. To find the y-intercept, set x equal to zero and solve for y.

Remember the graph of the straight line represents all possible pairs for which the sum invested is equal to \$50,000. In this example, the intercepts represent the extremes for the investments where all is put into one or the other. That is, all of the \$50,000 is put into stocks and nothing is put into mutual funds or vice versa.

Parallel and Perpendicular Lines

Figure 4-3 is an Excel plot showing three different investments where the money is invested in both mutual funds and the stock market. The line with x-intercept $(50,000, 0)$ and y-intercept $(0, 50,000)$ is the same as that given in Fig. 4-2. The line with x-intercept $(100,000, 0)$ and y-intercept $(0, 100,000)$ represents the possible investments of \$100,000 into stocks and mutual funds and has the equation $x + y = 100,000$. The line through the origin is the line $y = x$ and represents equal investments in stocks and mutual funds. The three lines may be written as $y = -x + 50,000$, $y = -x + 100,000$, and $y = x$. From Fig. 4-3, it can be seen that both lines $y = -x + 50,000$ and $y = -x + 100,000$ have equal slopes (both have slope $= -1$) and are parallel. One might say that these are parallel investments. The line $y = x$ has slope $m = 1$ and is seen to be perpendicular to the lines $y = -x + 50,000$ and $y = -x + 100,000$. The product of the slopes of the perpendicular lines is

$m_1 m_2 = -1$. This discussion leads to the following result concerning parallel and perpendicular lines.

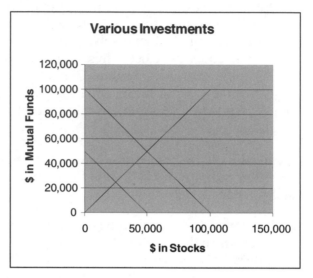

Fig. 4-3

Pattern

Suppose two lines have slopes m_1 and m_2. If $m_1 = m_2$, then the lines are parallel. If $m_1 m_2 = -1$, then the lines are perpendicular.

Linear Equations

Example 4-4: The last major league player to hit .400 was Ted Williams in 1941. George Brett came very close to this illusive mark in 1980. After 400 at bats, he was hitting exactly .400 with only 2 weeks left in the season. At that point in the season, the following question was asked: "If Brett comes to bat 50 more times this season, how many hits must he get to have a season average of .400?" The batting average for a player is defined as the number of base

hits divided by the number of times at bat. After 400 at bats, Brett had 160 hits, because 160/400 = .400. If x represents the number of hits needed in the remaining 50 at bats, then the following equation results:

$$(160 + x)/450 = .400$$

This problem represents a **linear equation.** A linear equation is an equation of the form

$$ax + b = c$$

where a, b, and c are numbers and x is a variable whose value is to be determined. The value of x that makes the equation true is called the **solution** of the linear equation.

The solution of a linear equation utilizes two basic principles, the **addition principle** and the **multiplication principle.** These principles are as follows:

Addition principle: If the same number is added to (or subtracted from) each side of any equation, an equivalent equation results.

Multiplication principle: If both sides of an equation are multiplied by (or divided by) the same nonzero number, an equivalent equation results. To see how these principles work, consider the solution to the above equation. First, both sides are multiplied by 450 to eliminate fractions.

$$450(160 + x)/450 = 450 (.400)$$

$$160 + x = 180$$

To complete the solution, 160 is subtracted from both sides to give $x = 20$. Brett would need 20 hits in his last 50 at bats to hit .400. As a historical baseball note, Brett ended the season with a .390 batting average. He won the batting title but missed the illusive .400 mark.

Example 4-5: Suppose $10,000 is invested in a mutual fund and no additional money is placed in the fund and no reduction in the fund occurs for a 1-year period. If the value of the fund at the end of that 1-year period is $11,500, what effective interest has the fund earned for that 1-year period? Suppose x represents the effective interest rate earned by the fund. Then the original balance plus the interest earned over the 1-year period must equal $11,500. This gives the following linear equation: $10,000 + x(10,000) = 11,500$. Using the addition principle, we have

$$10,000x = 11,500 - 10,000$$

$$10,000x = 1500$$

Using the multiplication principle for the equation $10,000x = 1500$, we divide both sides by 10,000 to obtain $x = 1500/10,000 = 0.15$ or 15%. The fund earned 15% for the past year.

Short Cuts

The linear equations encountered in this chapter are easy to solve. They involve the use of only four fundamental arithmetic operations. The solution of the two linear equations in this chapter by the use of MAPLE software is now illustrated. The MAPLE commands required to solve the equations are as follows.

```
> solve((160+x)/450=.400);          20.
> solve(10000+x*(10000)=11500);          3/20
```

To convert 3/20 to a percent, obtain the decimal equivalent and the multiply by 100 to obtain 15%.

The availability of software to solve more complex equations in algebra has proven to be invaluable to many individuals in quantitative professions.

Don't Forget

A *linear function* has an equation of the form $y = mx + b$.

If (x_1, y_1) and (x_2, y_2) represent any two points on a straight line, then the *slope* may be found by using the following equation:

$$m = \frac{\text{change in } y}{\text{change in } x} = \frac{y_2 - y_1}{x_2 - x_1} = \frac{y_1 - y_2}{x_1 - x_2}$$

The point where a line cuts the x-axis is called the *x-intercept* and the point where a line cuts the y-axis is called the *y-intercept*.

Suppose two lines have slopes m_1 and m_2. If $m_1 = m_2$, then the lines are parallel. If $m_1 m_2 = -1$, then the lines are perpendicular.

An equation of the form $ax + b = c$ is called a *linear equation*.

The solution of a linear equation is found by using the *addition* or *multiplication principle*.

The *addition principle* states that the same number may be added to or subtracted from each side of a linear equation to obtain an equivalent equation.

The *multiplication principle* states that if both sides of a linear equation are multiplied by or divided by the same nonzero number, an equivalent equation is obtained.

Questions

Test Yourself

1. Executive Car Rental charges $20 per day plus 10 cents per mile to rent one of its midsize cars. If y represents the daily cost and x represents the number of miles driven, express y in terms of x.

2. Using the linear function from Problem 1, find the daily cost for 100 and 200 miles driven. Then use these two points on the line to compute the slope of the line.

3. Find the x-intercept, the y-intercept, and the slope of the straight line given by the equation $y = 5x + 10$.

4. Use the slope to determine if the following pairs of lines are parallel, perpendicular, or niether:

   (a) $y = 5x - 2$ and $y = -0.2x + 3$.

   (b) $y = 12x - 6$ and $y = 12x + 6$.

   (c) $y = 2x + 7$ and $y = 3x - 7$.

5. Suppose you inherit $10,000 and decide to invest $4000 in mutual funds and the remainder in stocks. How many shares of stocks valued at $50 per share could you purchase? Let x represent the number of shares and give the linear equation that you must solve. What is the solution to this linear equation?

Answers

1. $y = 0.1x + 20$.

2. $y = 30$ when $x = 100$ and $y = 40$ when $x = 200$; slope = $(40 - 30)/(200 - 100) = 0.1$.

3. x-intercept = $(-2, 0)$, y-intercept = $(0, 10)$, and slope = 5.

4. (a) perpendicular;   (b) parallel;   (c) neither.

5. $50x + 4000 = 10,000$; solution is $x = 120$.

CHAPTER 5

Linear Inequalities

Do I Need to Read This Chapter?

➤ What is a linear inequality and what does the solution of a linear inequality mean?

➤ What are the addition and multiplication properties of inequalities?

➤ What is a linear inequality in two variables and what does the solution of a linear inequality in two variables mean?

➤ What is a graphical representation of the solution for a linear inequality in two variables and how can computer algebra software be used to find the graphical representation?

Linear Inequality in One Variable

Example 5-1: You can buy a box of floppy disks from a dealer on the Internet at Computer-supplies.com for $8 per box. The shipping cost is $10 per order, regardless of the number of boxes you order. You can buy the same box of floppies from a local computer store for $10.50 per box. How many boxes must you order before the cost of the purchase at Computer-supplies.com becomes less than from your local computer store? Consider first a solution that does not use algebra. Table 5-1 gives the cost for one through six boxes on both the Internet as well as from the local computer store. From the table, it is clear that when five or more boxes are purchased, it is cheaper to buy from the Internet dealer.

Table 5-1   Comparing Costs

NUMBER OF BOXES	COMPUTER-SUPPLIES.COM COST	COMPUTER STORE COST
1	$8 + 10 = 18$	10.50
2	$2(8) + 10 = 26$	$2(10.50) = 21$
3	$3(8) + 10 = 34$	$3(10.50) = 31.50$
4	$4(8) + 10 = 42$	$4(10.50) = 42$
5	$5(8) + 10 = 50$	$5(10.50) = 52.50$
6	$6(8) + 10 = 58$	$6(10.50) = 63$

An algebraic solution might proceed as follows. If we let x represent the number of boxes of floppies purchased, the cost for x boxes from the Internet dealer is $8x + 10$. The cost of x boxes from the local computer store is $10.50x$. We want to determine the value for x that makes the cost from the Internet dealer less than the cost from the local computer store. That is, we want to find the values of x for which the following is true:

$$8x + 10 < (\textit{is less than}) \ 10.50x$$

The symbol < is used in algebra to stand for "is less than" and the symbol > is used to represent "is greater than" when read from left to right. To solve this inequality, first we add $-8x$ to both sides of the inequality to obtain

$$8x + 10 - 8x < 10.50x - 8x \qquad \text{or} \qquad 10 < 2.50x$$

Now, both sides of the inequality $10 < 2.50x$ are divided by 2.50 to obtain $4 < x$, which is the same as the inequality $x > 4$.

$$10/2.50 < 2.50x/2.50 = x \qquad \text{or} \qquad 4 < x$$

We see that buying more than four boxes from the Internet dealer will result in a savings.

This introductory example introduces us to the following algebraic concepts. An inequality of the form $ax + b < cx + d$ or $ax + b > cx + d$ is called a **linear inequality.** For the linear inequality $8x + 10 < 10.50x$, we have $a = 8$, $b = 10$, $c = 10.50$, and $d = 0$. The set of all values of x for which the linear inequality is true is called the **solution set** for the inequality. The solution set for this inequality is $x > 4$.

Addition and Multiplication Properties of Inequalities

The **addition property of inequality** states that when the same number is added to or subtracted from both sides of an inequality, an equivalent inequality is obtained. Two inequalities are *equivalent* if they have the same solution set. The **multiplication property of inequality** states that both sides of an inequality may be multiplied or divided by the same positive number without changing the direction of the inequality, but that multiplying or dividing both sides by a negative number causes the reversal of the inequality symbol. Applying the multiplication property results in an equivalent inequality.

Danger!

Remember that when you multiply both sides of an inequality by the same positive number the sense of the inequality remains the same, but the sense is reversed if you multiply both sides by the same negative number.

Example 5-2: Suppose you are trying to decide between two car rental companies. Company A charges $50.00 per day with unlimited mileage. Company B charges $30.00 per day and 10 cents per mile. For a 5-day rental, how many miles would you need to travel so that choice of Company A would result in a cheaper rental? Table 5-2 illustrates how you might analyze the problem.

Table 5-2   When Is It Cheaper to Rent from Company A for a 5-Day Period?

COST TO RENT FROM COMPANY A	COST TO RENT FROM COMPANY B
5($50) = $250	5($30) + 0.10x = 150 + 0.10x
	Where x = the number of miles traveled

Company A is cheaper when $250 < 150 + 0.10x$. Subtracting 150 from each side we obtain $100 < 0.10x$.

$$250 - 150 < 150 + 0.10x - 150 \quad \text{or} \quad 100 < 0.10x$$

Dividing both sides by 0.10, we obtain $1000 < x$ or $x > 1000$.

$$100/0.10 < 0.10x/0.10 \quad \text{or} \quad 1000 < x$$

If you are likely to travel more than 1000 miles, it will be cheaper to rent from Company A. Note that the addition and multiplication principles were used to solve this problem.

Linear Inequality in Two Variables

Example 5-3: Suppose that you are not sure about the number of days you will need a rental car or how many miles you will travel. Now, rather than one variable involved, there are two variables. Suppose y represents the number of days that the car is rented and x represents the number of miles traveled. Table 5-3 shows how the problem may be analyzed.

Table 5-3   When Is It Cheaper to Rent from Company A?

COST TO RENT FROM COMPANY A	COST TO RENT FROM COMPANY B
$y(\$50) = \$50y$	$y(\$30) + 0.10x = 30y + 0.10x$
Where y = the number of days rented	Where x = the number of miles traveled

It is cheaper to rent from company A when $50y < 30y + 0.10x$. Subtracting $30y$ from both sides of the inequality, we obtain $20y < 0.10x$.

$$50y - 30y < 30y + 0.10x - 30y \qquad \text{or} \qquad 20y < 0.10x$$

Dividing both sides of this new equation by 20, we obtain $y < 0.005x$.

$$20y/20 < 0.10x/20 \qquad \text{or} \qquad y < 0.005x$$

The solution set for this inequality is best illustrated graphically. Figure 5-1 gives the solution to the problem.

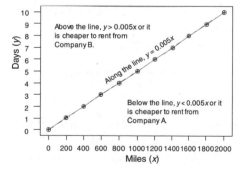

Fig. 5-1

The line $y = 0.005x$ divides the rectangle (which is in the first quadrant of the rectangular coordinate system) into two regions. For any point above the line $y = 0.005x$, $y > 0.005x$ and for any point below the line $y = 0.005x$, $y < 0.005x$.

Point A (400, 5) in Fig. 5-2, for example, is above the line $y = 0.005x$ and $y > 0.005x$ since $5 > 0.005(400) = 2$. Point B (1500, 5) is below the line $y = 0.005x$ and $y < 0.005x$ since $5 < 0.005(1500) = 7.5$.

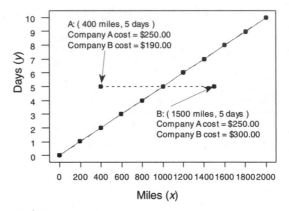

Fig. 5-2

The solution set consists of all points below the line $y = 0.005x$ in Fig. 5-1. This example illustrates a **linear inequality in two variables.** A linear inequality in two variables is an inequality containing two variables, each of which is raised to the first power. The **solution to a linear inequality in two variables** is a set of points in the rectangular coordinate system that makes the inequality true.

Short Cuts

If you are in a financial planning profession that occasionally requires the use of algebra to solve problems but do not wish to perform all the details to obtain the solution, software is often invaluable. Example 5-4 illustrates this.

Graphical Representation of the Solution Set for a Linear Inequality in Two Variables

Example 5-4: Suppose you are planning to invest in two different stocks. Stock A costs $60 per share and stock B costs $15 per share. You also wish to invest between $15,000 and $30,000. What are the different possibilities for the number of shares of stock A and stock B that may be purchased to satisfy the investment requirements? If we let x represent the number of shares of stock A to be purchased and y represent the number of shares of stock B to be purchased, then the cost for the total shares is $60x + 15y$. This total cost must be between $15,000 and $30,000. That is, $15,000 < 60x + 15y < 30,000$. MAPLE may be used to plot the set of all possible solutions to this inequality. The MAPLE commands are as follows.

```
inequal( { 15000<60*x+15*y,60*x+15*y<30000}, x=0..500,
  y=0..2000,
optionsfeasible=(color=gray),
optionsexcluded=(color=white),labels = ["Shares of
  A","Shares of B"]);
```

Figure 5-3 shows the solution as the shaded portion of the plot.

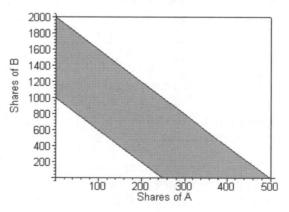

Fig. 5-3

Quick Tip

The region shown in Fig. 5-3 actually extends into both the second and fourth quadrants of the coordinate plane. However, because the number of shares of stock A must range between 0 and 500 and the number of shares of stock B must range between 0 and 2000, only the portion in the first quadrant is plotted.

If any point in the gray region is selected, it will satisfy the inequality $15{,}000 < 60x + 15y < 30{,}000$. For example, it is clear that the point $(300, 400)$ is in the shaded region and 300 shares of A and 400 shares of B costs $60(300) + 15(400) = \$24{,}000$ which satisfies the requirement that the investment be between \$15,000 and \$30,000. The line $60x + 15y = 15{,}000$ forms the lower boundary of the region and the line $60x + 15y = 30{,}000$ forms the upper boundary of the region.

Pattern

To solve a linear inequality of the form $ax + by < c$ or $ax + by > c$, first plot the line $ax + by = c$. This line divides the rectangular coordinate system (also called the xy plane) into two regions. For one region, $ax + by$ will be less than c for every point in the region, and for the other region, $ax + by$ will be greater than c for every point in the region. Choose any point on one side of the line and determine which of the two inequalities is true for this point. The same inequality will be true for every point on this side of the line.

Example 5-5: To be sure that we understand this procedure, suppose we wish to find the solution set for the linear inequality $4x - 3y < 12$. A plot of the line $4x - 3y = 12$ is produced by the MAPLE command `plot((4*x-12)/3,x=-10..10,y=-10..10);` The line is shown in Fig. 5-4.

Suppose we check the arbitrarily chosen point $(0,0)$ in the inequality $4x - 3y < 12$. Note that $4(0) - 3(0) = 0$ is less than 12. Therefore, for all points above the line $4x - 3y = 12$, we have the inequality $4x - 3y < 12$. We would shade the region above the line

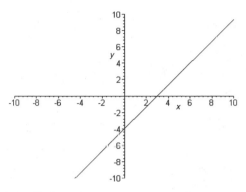

Fig. 5-4   **Fig. 5-5**

$4x - 3y = 12$ to illustrate our solution to the inequality $4x - 3y < 12$. This solution set is shown in Fig. 5-5.

Note that in both Figs. 5-4 and 5-5 the domain and range are shown from −10 to 10. The domain and range actually extend from minus infinity to plus infinity.

Don't Forget

A *linear inequality* in one variable is an inequality of the form $ax + b < cx + d$ or $ax + b > cx + d$.

The set of all x values for which a linear inequality is true is called the *solution set* for the inequality.

Two linear inequalities are *equivalent* if they have the same solution set.

The *addition property of inequality* states that when the same number is added to or subtracted from both sides of an inequality, an equivalent inequality is obtained.

The *multiplication property of inequality* states that both sides of an inequality may be multiplied or divided by the same positive number without changing the direction of the inequality, but that multiplying or dividing both sides by a negative number causes the reversal of the inequality symbol. An equivalent inequality results when the multiplication property is applied.

A *linear inequality in two variables* is an inequality containing two variables, each of which is raised to the first power.

The *solution to a linear inequality in two variables* is a set of points in the rectangular coordinate system that makes the inequality true.

Test Yourself

Questions

1. Solve the following linear inequality in one variable for x.
$$2x + 10 > x + 15$$

2. Solve the following linear inequality in one variable for x.
$$6x - 15 < 2x + 5$$

3. Solve the following linear inequality in the variables x and y for y.
$$2x + 7y > 10 - x + 3y$$

4. Solve the following linear inequality in the variables x and y for y.
$$3x + 6y + 14 > 10 + 2x + 3y$$

5. Sketch the solution to the following inequality graphically.
$$4 < x + y < 6, \text{ for } x \text{ and } y \text{ between } -10 \text{ and } 10$$

Answers

1. $x > 5$.

2. $x < 5$.

3. $y > (10 - 3x)/4$.

4. $y > (-x - 4)/3$.

5.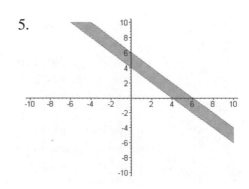

CHAPTER 6

Absolute Values

Do I Need to Read This Chapter?

→ What is the absolute value of a number?

→ What docs the graph of the absolute value function look like?

→ How do I solve equations involving the absolute value function?

→ How do I solve inequalities involving the absolute value function?

→ How can I use MAPLE to solve equations or inequalities that involve the absolute value function?

Signed Numbers

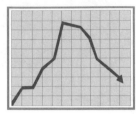

Table 6-1 gives information concerning the stocks for three companies listed in the New York Stock Exchange (NYSE).

Table 6-1   Stock Price Information for Three Stocks

52-WEEK HIGH	52-WEEK LOW	STOCK	LAST	CHANGE
76½	48⅜	AT&T	75	+⁷⁄₁₆
88¹⁵⁄₁₆	53⅜	Coca-Cola	68⅞	−⁹⁄₁₆
187¹⁵⁄₁₆	95⅝	IBM	180	+2¾

Example 6-1: Table 6-1 gives the high for the past year, the low for the past year, the name of the stock, the price per share at yesterday's closing, and the change from the day before. For Coca-Cola, the maximum price per share during the past year was $88\frac{15}{16} =$ $88.9375, the minimum during the past year was $53\frac{5}{8}$ = $53.625, the cost at closing yesterday was $68\frac{7}{8}$ = $68.875, and the cost per share the day before yesterday was $68\frac{7}{8} - (-\frac{9}{16}) = 68.875 + 0.5625 =$ $69.4375. The numbers in the column labeled Change are **signed numbers.** A thorough discussion of signed numbers is found in Chapter 1. A positive change means the stock increased in value from the previous day and a negative change means that the stock decreased in value from the previous day.

Absolute Value of a Signed Number

If we are told that the change in a stock was $2 and no sign or direction is given for the change, then we know that the **magnitude** or

absolute value of the change was $2, but we do not know if it went up or down in value. The absolute value of a number gives its magnitude but not its direction. If the change in the value of the stock is given as +2, we know that it went up by $2 and if the change is given as −2, we know that it went down $2. The absolute value of +2 is 2 and the absolute value of −2 is also 2. Remember the absolute value gives the magnitude but not the direction.

The absolute value of any quantity x is written as $|x|$ and is defined as follows:

$$|x| = \begin{cases} -x & \text{if} & x < 0 \\ x & \text{if} & x \geq 0 \end{cases}$$

Graph of the Absolute Value Function

Table 6-2 gives the changes in stock prices for several values between −1 and +1, along with their absolute values. Figure 6-1 is a Minitab plot of the points shown in Table 6-2. If all the values between −1 and +1 are filled in, the graph for $y = |x|$, shown in Fig. 6-2, is obtained. The MAPLE command `plot(abs(x),x=-1..1);` is used to obtain this graph.

Table 6-2 Some Changes in Stock Prices and Their Absolute Values

CHANGE	|CHANGE|	CHANGE	|CHANGE|
.125	.125	−.125	.125
.250	.250	−.250	.250
.375	.375	−.375	.375
.500	.500	−.500	.500
.625	.625	−.625	.625
.750	.750	−.750	.750
.875	.875	−.875	.875
1	1	−1	1

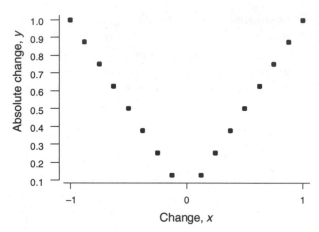

Fig. 6-1

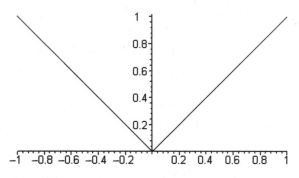

Fig. 6-2

Although the graph of the function $y = |x|$ in Fig. 6-2 is shown for the domain $-1 < x < 1$, the graph of $y = |x|$ for any domain would look similar to this graph. Many of the software packages are capable of constructing graphs of functions involving the absolute value function.

Solving Equations Containing the Absolute Value Function

Newspapers, television, and radio report the results of surveys and polls daily if not hourly.

Example 6-2: Suppose we are interested in the percentage of the population that favors campaign finance reform. A poll of 1000 individuals is taken and it is found that 510 or 51% in the poll are for campaign finance reform. The **margin of error** for the poll is reported to be 3%. This means that it is highly likely (a 95% chance) that the absolute difference between the unknown population percentage that favors campaign finance reform and the 51% found in the poll is less than 3%. To express this algebraically, suppose x represents the population percentage that favors campaign finance reform. Then we are very confident that the following inequality is true:

$$|x - 51\%| \leq 3\%$$

That is, the maximum difference between the poll result and the true population value is likely 3% or less. To find the lower and upper limits for x, we need to solve the following equation:

$$|x - 51\%| = 3\%$$

From the definition of absolute value, it follows that

$$x - 51\% = -3\% \quad \text{or} \quad x - 51\% = 3\%$$

Adding 51% to both sides of these equations, we find that

$$x = 48\% \quad \text{or} \quad x = 54\%$$

That is, the true population percentage that supports campaign finance reform is quite likely between 48 and 54%.

Example 6-3: Consider two golfers who shoot 85 on the average. However, one golfer is very consistent, whereas the other one is erratic. A statistical measure called the **average deviation** utilizes the absolute value function to compare the consistency of the two golfers. Sam scored 83, 82, 90, 84, and 86. The **mean** m for these five scores is 85. Table 6-3 illustrates how to compute absolute deviations from the mean. The symbol $\sum$ indicates that we are to add the quantities following the symbol.

Table 6-3   Computing Absolute Deviations

SCORE, x	DEVIATION $x - m$	ABSOLUTE DEVIATION $\lvert x - m \rvert$
83	$83 - 85 = -2$	2
82	$82 - 85 = -3$	3
90	$90 - 85 = 5$	5
84	$84 - 85 = -1$	1
86	$86 - 85 = 1$	1
	$\sum(x - m) = 0$	$\sum \lvert x - m \rvert = 12$

The **average deviation** is defined to be the mean of the five absolute deviations. That is, the **average deviation** for n numbers is $\sum \lvert x - m \rvert / n = \dfrac{12}{5} = 2.4$.

Short Cuts

This computation can be performed using the Microsoft Excel spreadsheet. If the five scores are put into cells A1 through A5 of the spreadsheet and the command =AVEDEV(A1:A5) is given, the value 2.4 is obtained.

John scored 70, 100, 85, 80, and 90 to obtain an average score of 85. If we use the Excel spreadsheet, it is found that the average deviation is 8. What can we conclude from the finding that Sam had an average deviation of 2.4 strokes and John had an average deviation of 8 strokes? Both golfers scored the same on the average. However, Sam is more consistent because he deviated from his mean by 2.4 strokes on the average, whereas John deviated from his mean by 8 strokes on the average. The average deviation is one of many measures of variation used by statisticians. The standard deviation, interquartile range, harmonic mean, and geometric mean are some other statistical measures of dispersion or variation.

Equations Involving Absolute Values

Example 6-4: Forty-five percent of those polled said that President Clinton was an effective leader during his first term as president. The margin of error for the poll was reported as 3%. Find the lower and upper limits for the percentage of the population that regard him as an effective leader. As in Example 6-2, we have the following equation involving the absolute value function:

$$|x - 45\%| = 3\%$$

It follows from the definition of absolute value that

$$x - 45\% = -3\% \qquad \text{or} \qquad x - 45\% = 3\%$$

Adding 45% to both sides of the equations, we obtain the following.

$$x = 42\% \qquad \text{or} \qquad x = 48\%$$

Therefore, we are very confident (not certain, but reasonably confident) that the percentage of the population that regard him as an effective leader is somewhere between 42 and 48%.

Quick Tip

The lower and upper limits for the population percentage are obtained by simply subtracting the margin of error from the poll percentage and adding the margin of error to the poll percentage, respectively.

Examples 6-2 and 6-4 introduce us to the solution of *absolute value equations.* An absolute value equation is one of the form $|ax + b| = c$, where a, b, and c are constants and a does not equal 0.

Pattern

Absolute value equations have no solutions or one or two solutions. If $c < 0$, the equation has no solutions; if $c = 0$, it has one solution; and if $c > 0$, it has two solutions as in Examples 6-2 and 6-4.

Short Cuts

The solution to the equation $|x - 45| = 3$ may also be found by the use of MAPLE. The solution is obtained by using the solve command as follows:

```
solve( abs(x - 45) = 3);
42, 48
```

The MAPLE solutions to the equations $|3x - 48| = 0$ and $|x - 65| = -3$ are as follows:

```
solve( abs(3*x - 48) = 0 );
16
solve( abs(x - 65) = -3 ); No answer is given, since it
   has no solutions.
```

Note that the solution given for $|3x - 48| = 0$ is 16. We see that this solution is correct since $|3(16) - 48| = |48 - 48| = 0$. Also, it is clear that $|x - 65| = -3$ has no solutions because the absolute value of anything is always zero or positive and therefore cannot equal -3.

Inequalities Involving Absolute Values

Table 6-2 gives some of the possible changes in stock prices between -1 and $+1$, along with their absolute values. Notice that whereas the changes range between -1 and $+1$, their absolute values range between 0 and $+1$. This implies that

$$|\text{change}| < 1 \qquad \text{is equivalent to} \qquad -1 < \text{change} < 1$$

In general, the following is true for an absolute value inequality of the form $|ax + b| < c$.

$$|ax + b| < c \qquad \text{is equivalent to} \qquad -c < ax + b < c$$

Example 6-5: In order that a pair of gears fit properly, it is necessary that the diameter of each gear deviate from 4 inches by less than 0.025 inches. That is, the diameter x needs to satisfy the following absolute inequality: $|x - 4| < 0.025$. We wish to find the acceptable range of diameters.

$$|x - 4| < 0.025 \qquad \text{is equivalent to} \qquad -0.025 < x - 4 < 0.025$$

Using the additive property of inequalities, discussed in Chapter 5, we add 4 to each part of the inequality to obtain the solution $3.975 < x < 4.025$. The solution using MAPLE is:

```
solve(abs(x-4)<0.025);
RealRange(Open(3.975000000), Open(4.025000000))
```

The solution means that all the numbers between 3.975 and 4.025 are included. The term Open means that the numbers 3.975 and

4.025 are not included, just the numbers in between them. If the inequality had been $|x - 4| \leq 0.025$, the MAPLE solution would be as follows. Note that the open is left off here, because the endpoints can be included.

```
Solve (abs (x-4)<=0.025);
Real Range (3.975000000, 4.025000000)
```

Gears with diameters less than 3.975 or diameters greater than 4.025 would not fit properly. This would correspond to x values for which $|x - 4| > 0.025$. This is equivalent to $x - 4 < -0.025$ *or* $x - 4 > 0.025$. To see that these are equivalent, solve each part separately as follows. Add 4 to both sides of the two expressions connected by *or* to obtain $x < 4 - 0.025 = 3.975$ or $x > 4 + 0.025 = 4.025$. In general, the following is true for an **absolute value inequality of the form $|ax + b| > c$.**

$$|ax + b| > c \qquad \text{is equivalent to} \qquad ax + b < -c \qquad \text{or} \qquad ax + b > c$$

Example 6-6: *USA Today* lists the leading gainers and the leading losers for the New York Stock Exchange (NYSE). Suppose, we want to list the stock whose absolute percent change exceeds 20%. We may express this using algebra as follows. Let x be the percent change for a stock listed in the NYSE. Then we are interested in those stocks for which $|x| > 20\%$. The solution to this absolute inequality is $x < -20\%$ or $x > 20\%$. Those stocks that showed a loss and for which the percent change exceeded 20% would be the **leading losers** and those that showed a gain for which the percent change exceeded 20% would be the **leading gainers.**

Short Cuts

The MAPLE solutions of $|x - 4| > 0.025$ and $|x| > 20\%$ are as follows:

```
solve( abs(x -4) > 0.025);
RealRange(Open(4.025000000), infinity), RealRange(-
   infinity, Open(3.975))
solve( abs(x) > 20);
RealRange(Open(20), infinity), RealRange(-infinity,
   Open(-20))
```

The MAPLE solution expressed as RealRange(Open(20), infinity) means that the leading gainers would consist of all those stocks with a percent gain of more than 20%. The term Open means we would not include one with exactly 20% gain. The solution RealRange(−infinity, Open(−20)) means the leading losers would be those with a 20% loss or more.

Danger!

A caution concerning the use of software is in order at this point. The MAPLE solution to the equation $|x-4| > 0.025$ is given as RealRange(−infinity, Open(3.975)) allows for negative solutions for x. But recall that x represents the diameters of the gears and cannot be negative. This solution should be modified to be (0, Open(3.975)) because negative diameters are not possible.

Don't Forget

Numbers written with a positive or a negative sign are called *signed numbers*. They indicate magnitude as well as direction.

The *absolute value of a number* is always equal to or greater than zero. The definition is given as follows:

$$|x| = \begin{cases} -x & \text{if} & x < 0 \\ x & \text{if} & x \geq 0 \end{cases}$$

The graph of $y = |x|$ is *V-shaped* like the ones shown in Figs. 6-2 and 6-3.

The *absolute value equation* $|ax + b| = c$ has no solutions if $c < 0$, one solution if $c = 0$, and two solutions if $c > 0$.

The *absolute value inequality* $|ax + b| < c$ is equivalent to $-c < ax + b < c$.

The *absolute value inequality* $|ax + b| > c$ is equivalent to $ax + b < -c$ or $ax + b > c$.

Test Yourself

Questions

1. For the IBM and AT&T stock information given in Table 6-1, find the cost per share for the day before yesterday.

2. The results of a poll recently reported that 80% of the students in grades 6 through 12 had experienced some form of bullying. The margin of error was reported as 3%. What are the statistical limits for the percentage of all students in grades 6 through 12 who have experienced some form of bullying?

3. Suppose your scores in four rounds of golf were 85, 90, 95, and 90. What is the average deviation for these four rounds?

4. Find all solutions to the following equations:

   (a) $|x + 4| = 1$.

   (b) $|x + 4| = 0$.

   (c) $|x + 4| = -1$.

5. Find all solutions to $|x + 4| < 1$.

6. Find all solutions to $|x + 4| > 1$.

Answers

1. For IBM, cost $= 177\frac{1}{4}$. For AT&T, cost $= 74\frac{9}{16}$.

2. 77 to 83%

3. 2.5

4. (a) -3 and -5 (b) -4 (c) No solutions

5. $-5 < x < -3$.

6. $x < -5$ or $x > -3$.

Systems of Linear Equations

System of Two Equations in Two Unknowns

Example 7-1: Suppose your car's radiator holds 12 quarts. You drain the system and want to mix your 100% antifreeze coolant with water to obtain a 60% antifreeze mixture, because the instructions indicate that this mixture will protect your engine down to 20 degrees below zero and that is all the protection you will need. You need to determine how much water and pure antifreeze to mix to form a 60% mixture. Represent the amount of water needed by x and the amount of pure antifreeze needed by y. Because the total capacity of the radiator is 12 quarts, $x + y$ must equal 12.

$$x + y = 12$$

In addition to this equation, we also know that the 12 quarts mixture must contain $12(0.60) = 7.2$ quarts of antifreeze because it is to be a 60% mixture. The 7.2 quarts must come from the y quarts of antifreeze, because the water does not supply any antifreeze.

$$y = 7.2$$

These two equations form a **system of equations** and we express the system as follows:

$$\begin{cases} x + y = 12 \\ \quad\;\; y = 7.2 \end{cases}$$

Solution to a System of Two Equations in Two Unknowns

The simplest method of solution to this **system of two equations in two unknowns** is the **method of substitution.** Simply substitute the value of y from the second equation into the first equation to obtain

$$x + 7.2 = 12$$

Now, subtract 7.2 from both sides to obtain

$$x = 12 - 7.2 = 4.8$$

Thus to obtain a 60% mixture of antifreeze, mix 7.2 quarts of pure antifreeze with 4.8 quarts of water. The **solution to the system of two equations in two unknowns is the point $(x, y) = (4.8, 7.2)$. A solution to a system of two equations in two unknowns is a pair of values for which *both* equations are true.**

Short Cuts

MAPLE can be used to solve systems of equations. The MAPLE command for the solution to the above system is as follows.

```
Solve ({x+y=12,y=7.2}, {x, y});
{y = 7.2, x = 4.8}
```

Graphical Solution to a System of Two Equations in Two Unknowns

A graphical solution of two linear equations in two unknowns is obtained by plotting the graphs of both equations on the same coordinate system and noting where the lines intersect. This is shown in Fig. 7-1. This MAPLE plot is produced by the following command.

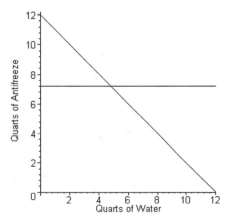

Fig. 7-1

```
plot([12-x,7.2],x=0..12,labels=["Quarts of Water",
  "Quarts of Antifreeze"]);
```

Notice that the plotted lines $x + y = 12$ and $y = 7.2$ intersect at the point $(4.8, 7.2)$.

The algebraic technique of solving two equations in two unknowns is an extremely useful technique for solving many real-world problems. This technique will now be illustrated by the use of several examples.

Example 7-2: Suppose you have just inherited $2500 and would like to invest some in stocks and some in mutual funds. A share of ACME Computer stock sells for $50 per share and a unit of Science/Technology mutual funds sells for $75 per unit. If you wish to purchase a total of 40 shares/units, how many shares of ACME and how many units of Science/Technology (S/T) can you buy with an investment of $2500? A solution not using algebra is shown in Table 7-1. Note from the table that the solution to the problem is to buy 20 shares of ACME Computer stock and 20 units of Science/Technology mutual funds. This will result in the total utilization of the $2500 and a total of 40 units/shares.

Table 7-1 Trial and Error Procedure to Find Solution to Investment Problem

SHARES OF ACME (x)	UNITS OF S/T (y)	$50x + 75y$ COST($)	SHARES OF ACME (x)	UNITS OF S/T (y)	$50x + 75y$ COST($)
0	40	3000	21	19	2475
1	39	2975	22	18	2450
2	38	2950	23	17	2425
3	37	2925	24	16	2400
4	36	2900	25	15	2375
5	35	2875	26	14	2350
6	34	2850	27	13	2325
7	33	2825	28	12	2300
8	32	2800	29	11	2275
9	31	2775	30	10	2250
10	30	2750	31	9	2225
11	29	2725	32	8	2200
12	28	2700	33	7	2175
13	27	2675	34	6	2150
14	26	2650	35	5	2125
15	25	2625	36	4	2100
16	24	2600	37	3	2075
17	23	2575	38	2	2050
18	22	2550	39	1	2025
19	21	2525	40	0	2000
20	**20**	**2500**			

The algebraic solution to this problem would proceed as follows. Suppose x and y are defined as in Table 7-1. Because a total of 40 shares is to be purchased,

$$x + y = 40$$

Also, because the total cost is $2500,

$$50x + 75y = 2500$$

We have the following system of two equations in two unknowns.

$$\begin{cases} x + y = 40 \\ 50x + 75y = 2500 \end{cases}$$

Addition/Subtraction Method for Solving a System of Equations

This system can be solved by the previously discussed methods of substitution or by graphing. You are encouraged to use both methods to solve this system. The addition/subtraction method to solve the system proceeds as follows. Suppose that we multiply the equation $x + y = 40$ on both sides by 50 to obtain

$$50(x + y) = 50(40) \quad \text{or} \quad 50x + 50y = 2000$$

Next we subtract this equation from the equation $50x + 75y = 2500$ as follows.

$$50x + 75y = 2500$$
$$\underline{-50x - 50y = 2000}$$
$$25y = 500$$

Solving for y, we have $y = 500/25 = 20$. Substituting this result into $x + y = 40$, we have

$$x + 20 = 40 \quad \text{or} \quad x = 20$$

This is the same result as obtained from Table 7-1.

Pattern

The addition/subtraction method for solving simultaneous equations is summarized as follows: Multiply one or both of the equations by suitable numbers to obtain the same absolute value for the coefficient of either variable. Then add or subtract the equations to eliminate one of the variables. Solve the resulting equation that has only one variable for that variable. Substitute the value obtained into either of the original equations and solve for the remaining variable.

This is one of many situations in algebra where an example is worth a thousand words. The discussion in the context of an example is much easier to understand than trying to understand a verbal discussion of the method.

In all the examples discussed so far in this chapter, the system of linear equations has a unique solution. That is, there is only one point, (x, y), for which both equations are true. Or, if both lines are plotted on a coordinate system, they are found to pass through a common point as in Fig. 7-1. This common point is the solution to the system. When the system has a single unique solution, the system of equations is said to be *consistent*. Sometimes a system of equations has no solutions. Such a system is said to be *inconsistent*. When a system has an infinite number of solutions, the system is said to be *dependent*. Although inconsistent or dependent systems rarely occur when using algebra to solve real-world problems, they are discussed because it is possible that such systems might be encountered. An example of each will now be discussed.

Example 7-3: Consider the following system: $\begin{cases} x - y = 50 \\ 5x - 5y = 250 \end{cases}$. A plot of this system is shown in Fig. 7-2.

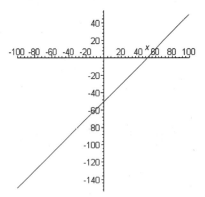

Fig. 7-2

That's right, when the two lines are plotted, it is found that they are in fact the same set of points. So any solution to one is automatically a solution to the other. The point with $x = 250$ and $y = 200$ is a solution since $250 - 200 = 50$. But it is a solution to the other equation also, since $5(250) - 5(200) = 1250 - 1000 = 250$. The point $(75, 25)$ is a solution to both equations also, since $75 - 25 = 50$ and $5(75) - 5(25) = 250$. There are an infinite number of solutions to this system.

Suppose we use the addition/subtraction method to solve the system and therefore multiply the equation $x - y = 50$ by 5 to obtain $5x - 5y = 250$ and subtract it from the other equation, we obtain

$$5x - 5y = 250$$
$$-5x + 5y = -250$$
$$0 = 0$$

Quick Tip

If both variables are eliminated when a system of linear equations is solved, there are infinitely many solutions if the resulting statement ($0 = 0$) is true. We call such a system a *dependent system.*

Example 7-4: Next, consider the system $\begin{cases} 4x - 2y = 100 \\ 2x - y = 25 \end{cases}$. A plot of this system is shown in Fig. 7-3.

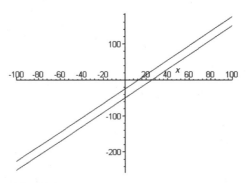

Fig. 7-3

In this case, the lines are parallel and have no points in common. There are no solutions. If the equation $2x - y = 25$ is multiplied by 2, we obtain $4x - 2y = 50$. When this equation is subtracted from the equation $4x - 2y = 100$, we obtain $0 = 50$ as follows.

$$4x - 2y = 100$$
$$4x - 2y = 50$$
$$0 = 50$$

Quick Tip

If both variables are eliminated when a system of linear equations is solved, there is no solution if the resulting statement (such as $0 = 50$) is false. We call such a system *an inconsistent system.*

Linear Programming

Example 7-5: In her financial portfolio, Lana plans to invest up to $30,000 in corporate and municipal bonds combined. The corporate bond investment requires at least $5000 and Lana does not wish to invest more than $10,000 in corporate bonds. She wants to invest between $10,000 and $25,000 in municipal bonds. The corporate bonds pay 6.5% simple interest per year and the municipal bonds pay 5.5% simple interest per year. What investment allocation will maximize her interest income as well as satisfy the stated constraints?

If the amount invested in corporate bonds (in thousands) is represented by x and the amount invested in municipal bonds (in thousands) by y, then the above discussion leads to the following constraints.

$$x + y \leq 30$$

$$5 \leq x \leq 10$$

$$10 \leq y \leq 25$$

Short Cuts

The region described by these inequalities may be plotted on graph paper or by the using the following MAPLE statements. The region is shown in gray in Fig. 7-4.

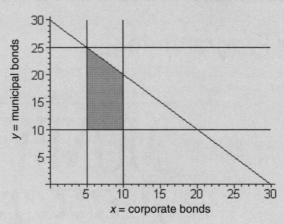

Fig. 7-4

```
> with(plots);
> inequal( { x+y<=30,x>=5,x<=10,y>=10,y<=25}, x=0..30,
  y=0..30,
> optionsfeasible=(color=gray),
> optionsexcluded=(color=white),labels=["x=corporate
  bonds","y=municipal bonds"]);
```

The interest earned (in thousands) for any point in this region is given by the expression

$$I = 0.065x + 0.055y$$

The function I is called the *objective function.* The corners of the region are listed in Table 7-2. We wish to find the maximum value of I for the points in the region of possible solutions. An area of mathematics called *linear programming* may provide an answer to this optimization problem. *It can be shown that under certain conditions, the objective function takes on its maximum and minimum values at the vertices or corners of the region.* The vertices of the region of possible solutions and the value of the objective function at these points are shown in Table 7-2. The maximum interest income is $1750 and it occurs when $10,000 is invested in corporate bonds and $20,000 is invested in municipal bonds. The minimum interest income is $875 and it occurs when $5000 is invested in corporate bonds and $10,000 in municipal bonds.

Table 7-2 Corners of the Region of Possible Solutions

CORNER	$I = 0.065x + 0.055y$
(5,10)	0.875 thousand
(5,25)	1.700 thousand
(10,20)	1.750 thousand
(10,10)	1.200 thousand

If the interest rates for the two types of bonds were reversed, Table 7-3 gives the values of the objective function at the corners of the polygon in

Table 7-3 Corners of the Region of Possible Solutions

CORNER	$I = 0.055x + 0.065y$
(5,10)	0.925 thousand
(5,25)	1.900 thousand
(10,20)	1.850 thousand
(10,10)	1.200 thousand

Fig. 7-4. The maximum now occurs when $5000 is invested in corporate bonds and $25,000 is invested in municipal bonds.

Quick Tip

Don't forget that in many linear programming problems, the optimum solution, subject to the stated constraints, occurs at a corner of the solution region. This is true if the solution region is a convex polygon, that is a polygon with no indentations.

System of Three Equations in Three Unknowns

Example 7-6: Books Galore sells books and has outlets across the country. Their advertising budget equals $100 million. The budgeted amount is divided among TV advertising, newspaper advertising, and website expenses on the Internet. The marketing department suggests that the amount spent on Internet expenses exceed twice the sum spent on TV and newspaper advertising by $10 million. In addition, the amount spent on TV advertising should be twice that spent on newspaper advertising. In order to determine how to divide the $100 million to accomplish the recommendations of the marketing department, we define the following variables.

x = the amount to be spent on newspaper advertising

y = the amount to be spent on TV advertising

z = the amount to be spent on website expenses on the Internet

The fact that the $100 million is to be divided among the three sources is expressed as

$$x + y + z = 100$$

The requirement that the amount spent on Internet expenses exceed twice the sum spent on TV and newspaper advertising by $10 million is expressed as

$$z = 2(x + y) + 10 \quad \text{or} \quad -2x - 2y + z = 10$$

The requirement that the amount spent on TV advertising should be twice that spent on newspaper advertising is expressed as

$$y = 2x$$

The system of three equations in three unknowns is expressed as follows:

$$\begin{cases} x + y + z = 100 \\ -2x - 2y + z = 10 \\ y = 2x \end{cases}$$

To solve this system, first we substitute $2x$ for y in both the first and second equation in the system to obtain the following system of two equations in two unknowns.

$$\begin{cases} 3x + z = 100 \\ -6x + z = 10 \end{cases}$$

To solve this reduced system, subtract the second equation from the first to obtain:

$$3x + z = 100$$
$$\underline{6x - z = -10}$$
$$9x = 90 \quad \text{or} \quad x = \$10 \text{ million}$$

Substitute $x = 10$ into $y = 2x$ to obtain $y = 2(10) = \$20$ million.

Finally, substituting these values in the equation $x + y + z = 100$, and then solving for z, we find

$$z = 100 - 10 - 20 = 70$$

It is easy to see that the solution $x = 10$, $y = 20$, and $z = 70$ satisfies the system of three equation in three unknowns.

Short Cuts

The MAPLE solution to this system proceeds as follows.

```
> solve( {x+y+z=100, -2*x-2*y+z=10, y = 2*x},{x,y,z} );
{x = 10, z = 70, y = 20}
```

Systems of Equations and Multiple Regression

A statistical technique called multiple regression is often used to model real-world phenomena. Blood pressure might be modeled by its relationship to other variables such as age, weight, type A personality measure, potassium to sodium ratio, and hours spent per week in contemplative meditation. These models require that systems of linear equations be solved. The blood pressure example would require that after data relating blood pressure and the other five variables were collected that a system of six equations in six unknowns be solved. This can be extremely challenging if performed by hand. Fortunately, such problems are solved by the use of software today. To see the power of such software consider the system of equations given in Example 7-7.

Example 7-7: Consider the following MAPLE solution to the given system of equations.

$$\begin{cases} a + b + c + d + e = 50 \\ a + b + c - 2d + e = 5 \\ 9a + b + c + d - e = 10 \\ a - b - c - d + e = -10 \\ 3a + 2b - 3c + 4d - 6e = -80 \end{cases}$$

```
> solve({a+b+c+d+e=50,a+b+c-2*d+e=5,9*a+b+c+d-e=10,a-b-
  c-d+e=-10,3*a+2*b-3*c+4*d-6*e=-80},{a,b,c,d,e});
{d = 15, e = 20, c = 10, b = 5, a = 0}
```

You can check the solution and verify that it satisfies all five equations. If you have several hours with nothing to do, you could try solving the system by elimination and gain a true appreciation for algebra software.

Don't Forget

If we consider two linear equations together, they form a *system of linear equations.*

A point (*a, b*) is a *solution* to a system of linear equations if the point (*a,b*) satisfies both equations.

In the *substitution method* for solving a system of linear equations, one of the equations is solved for *x* or *y*, and then the value is substituted into the other equation. The resulting equation in one variable is then solved and that value is substituted into either original equation. Another equation in one variable is obtained and the solution is then completed.

In the *graphical method* for solving a system of linear equations, the two straight lines are plotted on a rectangular coordinate system and the point of intersection of the two lines is the solution to the system.

In the *addition/subtraction method* for solving a system of linear equations, one or both of the equations are multiplied by suitable numbers to obtain the same absolute value for the coefficient of one of the variables. Then add or subtract the equations to eliminate one of the variables. Solve the resulting equation that has only one variable for that variable. Substitute the value obtained into either of the original equations and solve for the remaining variable.

If a system of linear equations has no solutions, it is said to be an *inconsistent system.* If the system has only one solution, it is said to be a *consistent system.* If the system has an infinite number of solutions, it is said to be a *dependent system.*

The graph of an inconsistent system of linear equations consists of two parallel lines, the graph of a consistent system consists of two lines that meet in a single point, and the graph of a dependent system of linear equations is a single line.

Linear programming techniques are used to maximize or minimize some objective function over a region that is determined by a system of multiple inequalities.

Test Yourself

Questions

1. A company plans to spend $150,000 on advertising expenses associated with radio, television, and newspaper commercials. Their total budget amount for advertising is $225,000. Their only other source of advertising is the Internet. Let x represent the amount allocated to radio, television, and newspaper commercials and let y represent the amount to be spent for Internet advertising. Express the above scenario in the form of two equations in two unknowns and give the solution to the system.

2. How many ounces of 30% acid solution and 90% acid solution must be mixed to obtain 50 ounces of 50% acid solution? Let x represent the amount of 30% acid solution and let y represent the amount of 90% acid solution. Give the two equations that describe this scenario and the solution to the system.

3. Classify each of the following systems as consistent, inconsistent, or dependent. For the consistent system, give the solution.

   (a) $\begin{cases} 2x - 4y = 13. \\ 4x - 8y = 26. \end{cases}$

   (b) $\begin{cases} 2x - 4y = 13. \\ 4x - 8y = 17. \end{cases}$

   (c) $\begin{cases} x + 2y = 12. \\ x - 3y = -13. \end{cases}$

4. ACE manufacturing produces music compact discs as well as data compact discs. A shipment of music compact discs requires 2 hours on machine A and 3 hours on machine B. A shipment of

data compact discs requires 3 hours on machine A and 1 hour on machine B. During a given week, machine A can be used to make these products in, at most, 24 hours and machine B can be used to make these products in, at most, 15 hours. The company makes profits of $4 thousand on a box of music compact discs and $3 thousand on a box of data compact discs. How many of each product must be produced to maximize weekly profits?

5. Find the solution to the following system of three equations in three unknowns.

$$\begin{cases} x + y + z = 60 \\ 2x - y + z = 30 \\ x + y - z = 0 \end{cases}$$

6. Verify that $a = 5, b = 10, c = 15, d = 20$ is a solution to the following system of four equations in four unknowns.

$$\begin{cases} a - b - c - d = -40 \\ a + 2b + d = 45 \\ a + b - c - d = -20 \\ a + 2b + 3c + 4d = 150 \end{cases}$$

Answers

1. $\begin{cases} x = 150{,}000. \\ x + y = 225{,}000. \end{cases}$

     $x = 150{,}000$ and $y = 75{,}000.$

2. $\begin{cases} x + y = 50. \\ 0.3x + 0.9y = 25. \end{cases}$

     $x = 33.3$ and $y = 16.7.$

3. (a) dependent;   (b) inconsistent;   (c) consistent, $x = 2, y = 5$.

4. Three shipments of music compact discs, six shipments of data compact discs, profit = $30,000.

5. $x = 10, y = 20,$ and $z = 30.$

CHAPTER 8

Polynomials

Do I Need
to Read
This Chapter?

➜ What is a polynomial?

➜ How do I add, subtract, multiply, and divide polynomials?

➜ How do solve polynomial equations?

➜ How can I use MAPLE to solve more complicated polynomial equations?

➜ How can I use MAPLE, Minitab, and Excel to graph polynomials?

Polynomial in One Variable

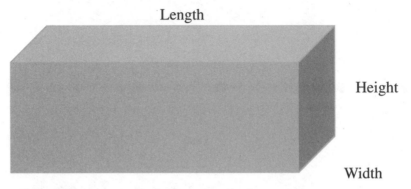

CustomBoxes.com, the largest source of custommade boxes on the Internet

Example 8-1: Custom Boxes Inc. manufactures custom-made boxes. One customer requires that the length equal the width plus 5 and that the height equal the width plus 2, for several different widths. If we let x represent the width, then the length needs to equal $x + 5$, and the height needs to equal $x + 2$. Summarizing, we have the following expressions for the width, length, and height.

$$\text{width} = x \qquad \text{length} = x + 5 \qquad \text{and} \qquad \text{height} = x + 2$$

Because the volume of the box is width times length times height, a formula for the volume V is

$$V = (x)(x + 5)(x + 2) = x^3 + 7x^2 + 10x$$

This expression for the volume is arrived at in the following way. First $(x + 5)(x + 2)$ is found using the FOIL method that is described in Chapter 1.

$$(x + 5)(x + 2) = x^2 + 7x + 10$$

Now the expression $x^2 + 7x + 10$ is multiplied by x to obtain the volume.

$$V = x^3 + 7x^2 + 10x$$

Table 8-1 gives information concerning some of the most common boxes made for this customer.

Table 8-1 Some of the Boxes Made by Custom Boxes Inc.

WIDTH (x)	LENGTH ($x + 5$)	HEIGHT ($x + 2$)	VOLUME	$x^3 + 7x^2 + 10x$
2	7	4	$2(7)(4) = 56$	56
4	9	6	$4(9)(6) = 216$	216
6	11	8	$6(11)(8) = 528$	528
8	13	10	$8(13)(10) = 1040$	1040
10	15	12	$10(15)(12) = 1800$	1800

Note that in Table 8-1 the volume can be found by multiplying the width times the length times the height, or by using the equation $V = x^3 + 7x^2 + 10x$, where x is the width of the box. The volume of the $2 \times 7 \times 4$ box is $2(7)(4) = 56$ or $(2)^3 + 7(2)^2 + 10(2) = 8 + 28 + 20 = 56$. This is an example of a **third-degree polynomial in one variable.** The letter x is used to represent the width of a given box and is called the **variable** in the polynomial. This polynomial has three *terms.* A **term** is a number or the product of a number and a variable raised to a whole-number power. The numerical part of a term is called the **coefficient.** The general definition of a **polynomial** is a term or some combination of sums and/or differences of terms. Polynomials will never contain terms with variables having negative exponents and there will never be variables in the denominator of a term. A polynomial with only one variable is called a **polynomial in one variable.** A polynomial with more than one variable is called a **polynomial in several variables.** We shall discuss only polynomials in one variable.

A polynomial containing one term is called a **monomial.** A polynomial containing two terms is called a **binomial.** A polynomial containing three terms is called a **trinomial.** The **degree of a term** in one variable is the exponent of the variable. The **degree of a polynomial** is the largest degree of any term in the polynomial. Table 8-2 gives some laws of exponents that apply when working with polynomials.

Table 8-2   Some Laws of Exponents

$x \neq 0$ AND a AND b INTEGERS	
$x^a \cdot x^b$	x^{a+b}
$(x^a)^b$	x^{ab}
x^a / x^b	x^{a-b}
x^{-a}	$1/x^a$
x^0	1

The expression for the volume of the boxes given earlier uses the laws of exponents given in Table 8-2 as well as the distributive law.

Addition and Subtraction of Polynomials

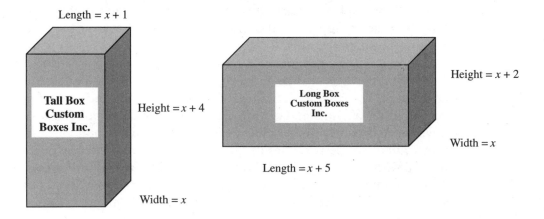

Length = $x + 1$

Tall Box
Custom
Boxes Inc.

Height = $x + 4$

Width = x

Long Box
Custom Boxes
Inc.

Height = $x + 2$

Width = x

Length = $x + 5$

Example 8-2: Suppose that Custom Boxes Inc. also produces another custom box with dimensions $x \times (x + 1) \times (x + 4)$. The volume of this tall box is given by $x(x + 1)(x + 4) = x^3 + 5x^2 + 4x$. As we saw earlier, the volume of the long box is $x^3 + 7x^2 + 10x$. The long box has more volume than the tall box. The difference is $(x^3 + 7x^2 + 10x) - (x^3 + 5x^2 + 4x)$. The difference is found as follows:

$$\begin{array}{r} x^3 + 7x^2 + 10x \\ \underline{-x^3 - 5x^2 - 4x} \\ 2x^2 + 6x \end{array}$$

The total volume of a long box plus a tall box is found by adding the two volumes. The sum is $(x^3 + 7x^2 + 10x) + (x^3 + 5x^2 + 4x)$. The sum is found as follows:

$$\begin{array}{r} x^3 + 7x^2 + 10x \\ \underline{x^3 + 5x^2 + 4x} \\ 2x^3 + 12x^2 + 14x \end{array}$$

This discussion introduces us to the **addition and subtraction of polynomials.** Table 8-3 gives the volumes for five different tall and five corresponding long boxes having the same width as well as their sums and differences. Note that the expressions $2x^2 + 6x$ and $2x^3 + 12x^2 + 14x$ do indeed give the difference and sum of the volumes.

Pattern

To add two polynomials, add like terms. In Example 8-2, the like terms x^3 and x^3 were added to obtain $2x^3$, the like terms $7x^2$ and $5x^2$ were added to obtain $12x^2$, and the like terms $10x$ and $4x$ were added to obtain $14x$. The result is then written as $2x^3 + 12x^2 + 14x$. To subtract two polynomials, change the signs of the terms in the polynomial being subtracted and then add like terms.

Table 8-3 Sums and Differences of Polynomials

WIDTH	LONG BOX VOLUME	TALL BOX VOLUME	DIFFERENCE	$2x^2 + 6x$	SUM	$2x^3 + 12x^2 + 14x$
2	$2(7)(4) = 56$	$2(3)(6) = 36$	20	20	92	92
4	$4(9)(6) = 216$	$4(5)(8) = 160$	56	56	376	376
6	$6(11)(8) = 528$	$6(7)(10) = 420$	108	108	948	948
8	$8(13)(10) = 1040$	$8(9)(12) = 864$	176	176	1904	1904
10	$10(15)(12) = 1800$	$10(11)(14) = 1540$	260	260	3340	3340

Quick Tip

When concepts like polynomials are discussed in most algebra books, they are treated in a rather abstract way. If the algebra student will keep in mind that polynomials actually represent some real-world entity such as the volume of a box, it makes the concept easier to understand. The sum of two polynomials represents the total volume of two boxes and the difference in two polynomials represents how much more volume one box has than another.

Multiplication of Polynomials

Example 8-3: Usedcars.com is an Internet location that sells used cars. Table 8-4 gives inventory information for this dealer. The first row of the table tells us that Usedcars.com has 19.6(100) = 1960 cars that are 1 year old. The average cost is 10.05(1000) = $10,050 per car for the 1-year-old cars. The total value of all the 1-year-old cars is 196.98(100,000) or $19,698,000. The rows for cars 2, 3, 4, and 5 years old are interpreted similarly.

Table 8-4   Inventory Information for Usedcars.com

AGE (YEARS)	NUMBER (100s)	AVERAGE VALUE ($1000s)	TOTAL VALUE ($100,000)
1	19.6	10.05	19.6(10.05) = 196.98
2	15.8	8.20	15.8(8.20) = 129.56
3	12.2	6.45	12.2(6.45) = 78.69
4	9.4	4.80	9.4(4.80) = 45.12
5	8.0	3.25	8.0(3.25) = 26.00

In this table, if the age column is represented by x, then the number of cars column is related to x by the following polynomial:

$$\text{Number} = 0.1x^3 - 0.5x^2 - 3.00x + 23.0$$

The average value column is related to x by the following polynomial:

$$\text{Average value} = 0.05x^2 - 2x + 12$$

To see clearly what this means, substitute $x = 1$ in the polynomial that represents number and note that you obtain 19.6, the value given in the table.

$$\text{Number} = 0.1(1)^3 - 0.5(1)^2 - 3.00(1) + 23.0 = 19.6$$

Verify the values in the Number column of Table 8-4 for $x = 2, 3, 4$, and 5 substituted into the polynomial.

Substitute $x = 4$ in the polynomial that represents average value to obtain the following:

$$\text{Average value} = 0.05(4)^2 - 2(4) + 12 = 4.80$$

The Total Value column should equal to the product of these two polynomials. That is,

$$\text{Total value} = (0.1x^3 - 0.5x^2 - 3.00x + 23.0)(0.05x^2 - 2x + 12)$$

We shall now show how to multiply these two polynomials.

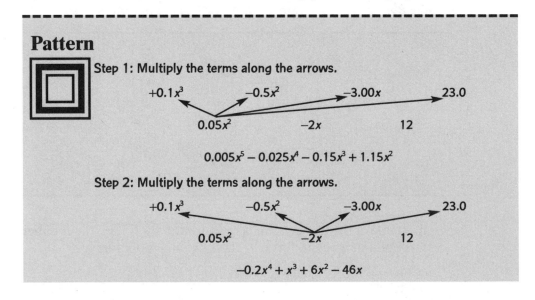

Pattern

Step 1: Multiply the terms along the arrows.

$+0.1x^3$ $-0.5x^2$ $-3.00x$ 23.0

$0.05x^2$ $-2x$ 12

$$0.005x^5 - 0.025x^4 - 0.15x^3 + 1.15x^2$$

Step 2: Multiply the terms along the arrows.

$+0.1x^3$ $-0.5x^2$ $-3.00x$ 23.0

$0.05x^2$ $-2x$ 12

$$-0.2x^4 + x^3 + 6x^2 - 46x$$

Step 3: Multiply the terms along the arrows.

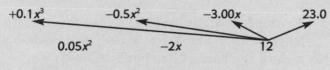

$$1.2x^3 - 6x^2 - 36x + 276$$

Step 4: Add the expressions shown in steps 1, 2, and 3 and then combine like terms.

$$0.005x^5 - 0.025x^4 - 0.15x^3 + 1.15x^2 - 0.2x^4 + x^3 + 6x^2 - 46x$$
$$+ 1.2x^3 - 6x^2 \quad 36x + 276$$

$$\text{Total value} = 0.005x^5 - 0.225x^4 + 2.05x^3 + 1.15x^2 - 82x + 276$$

Table 8-5 shows the polynomial representation of Table 8-4. If the *x* value from the Age column is substituted into the polynomials given in columns 2, 3, and 4, the numerical values are obtained. Table 8-5 is the polynomial representation of Table 8-4.

Table 8-5 Polynomial Representation for Table 8-4

AGE, x	NUMBER = $0.1x^3 - 0.5x^2$ $- 3.00x + 23.0$	AVERAGE VALUE = $0.05x^2$ $- 2x + 12$	TOTAL VALUE = $0.005x^5 - 0.225x^4 + 2.05x^3$ $+ 1.15x^2 - 82x + 276$
1	19.6	10.05	196.98
2	15.8	8.20	129.56
3	12.2	6.45	78.69
4	9.4	4.80	45.12
5	8.0	3.25	26.00

Quick Tip

In the discussion of polynomial multiplication, we saw that the numbers of cars of different ages could be represented as a polynomial, the average values could be represented by a polynomial, and the total value could be represented by the product of these two polynomials.

Division of Polynomials

Example 8-4: In Example 8-1, we found that a box with width $= x$, length $= x + 5$, and height $= x + 2$, had volume $= x^3 + 7x^2 + 10x$. The area of the base of this box is $x(x + 5) = x^2 + 5x$. If the volume is divided by the height, the area of the base is obtained. That is,

$$(x^3 + 7x^2 + 10x)/(x + 2) = x^2 + 5x$$

This is now demonstrated by the use of polynomial division.

Step 1: Divide the first term of the dividend by the first term of the quotient to obtain x^2. That is, $\dfrac{x^3}{x} = x^2$.

$$x + 2 \overline{)x^3 + 7x^2 + 10x} \quad \overset{x^2}{}$$

Step 2: Multiply the entire divisor by the quotient. That is, $x^2(x + 2) = x^3 + 2x^2$

$$
\begin{array}{r}
x^2 \\
x + 2 \overline{)x^3 + 7x^2 + 10x} \\
\underline{x^3 + 2x^2 }
\end{array}
$$

Step 3: Subtract this product from the dividend. To subtract a polynomial, add its negative and then bring down the next term from the dividend.

$$
\begin{array}{r}
x^2 \\
x + 2 \overline{)x^3 + 7x^2 + 10x} \\
\underline{x^3 + 2x^2 } \\
5x^2 + 10x
\end{array}
$$

Step 4: Divide the first term of the new dividend by the first term of the divisor to obtain $5x$. That is, $\dfrac{5x^2}{x} = 5x$.

$$
\begin{array}{r}
x^2 + 5x \\
x + 2 \overline{)x^3 + 7x^2 + 10x} \\
\underline{x^3 + 2x^2 } \\
5x^2 + 10x
\end{array}
$$

Step 5: Multiply the entire divisor by the second term of the quotient.

$$x^2 + 5x$$
$$x + 2\overline{)x^3 + 7x^2 + 10x}$$
$$\underline{x^3 + 2x^2}$$
$$5x^2 + 10x$$
$$5x^2 + 10x$$

Step 6: Subtract this product from the new dividend.

$$x^2 + 5x$$
$$x + 2\overline{)x^3 + 7x^2 + 10x}$$
$$\underline{x^3 + 2x^2}$$
$$5x^2 + 10x$$
$$\underline{5x^2 + 10x}$$
$$\text{Remainder} = 0$$

Short Cuts

The power and speed of computer algebra software packages is seen clearly when used to perform polynomial division. Consider the following MAPLE solution to Example 8-4. The command simplify may be used to perform the polynomial division.

```
> simplify((x^3 + 7*x^2 + 10*x)/(x + 2));
x² + 5x
```

Polynomial Equations

Example 8-5: The area of the base of the box in Example 8-4 is $x^2 + 5x$. Suppose the customer requires a box with base area equal to 6 square units. This requirement results in the following equation:

$$x^2 + 5x = 6 \qquad \text{or} \qquad x^2 + 5x - 6 = 0$$

This equation is referred to as a **second-degree polynomial equation in x.** There are two values of x for which this equation is true.

They are called the **roots of the polynomial.** One approach to solving this equation is to try to express $x^2 + 5x - 6$ in **factored form.**

Using the factoring technique discussed in Chapter 1, we find that

$$x^2 + 5x - 6 = (x - 1)(x + 6)$$

Our second-degree polynomial $x^2 + 5x - 6 = 0$ may be replaced by

$$(x - 1)(x + 6) = 0$$

This means that either the factor $x - 1 = 0$ or the factor $x + 6 = 0$. Setting these factors equal to zero and solving for x we find that $x = -6$ or 1. Since x represents the width and must be a positive measurement, the solution is $x = 1$.

Example 8-6: In Example 8-1 a custom-made box with width = x, length = $x + 5$, height = $x + 2$, and volume = $x^3 + 7x^2 + 10x$ was discussed. Suppose we also require that the volume of the box be equal to 56 cubic units. What dimensions would be needed to meet this requirement? This requirement gives the following equation.

$$x^3 + 7x^2 + 10x = 56 \qquad \text{or} \qquad x^3 + 7x^2 + 10x - 56 = 0$$

This equation is called a **third-degree polynomial equation in x.** It is much more difficult to express a third-degree polynomial in factored form than a second-degree polynomial equation. The algebraic methods for finding the roots for such polynomials are beyond the scope of this book.

Short Cuts

However, the solutions are easily found using computer algebra software. The solution using MAPLE is as follows.

```
> solve(x^3 + 7*x^2 + 10*x - 56) = 0;
(2, -9/2 + 1/2 I sqrt(31), -9/2 - 1/2 I sqrt(31))
```

Three solutions are indicated. The real number 2 and two complex numbers. Complex numbers are discussed in Chapter 10. To see that $x = 2$ is a solution, note that

$$(2)^3 + 7(2)^2 + 10(2) - 56 = 8 + 28 + 20 - 56 = 0$$

Graphs of Polynomials

Polynomials are often used to model real-world phenomena. In this chapter, we have considered examples in which polynomials were used to model area and volume as a function of width, the number of used cars as a function of age, and the price of used cars as a function of age. The list of possible polynomial models is endless. The graph of a polynomial model is very useful in understanding real-world phenomena. Computer software packages are extremely useful in graphing polynomials. The following example gives the graph of a polynomial produced by using Minitab, Excel, and MAPLE.

Example 8-7: Suppose we wish to plot the polynomial $y = x^3 + 7x^2 + 10x - 56$ for values of x between -3 and 3. Figs. 8-1, 8-2, and 8-3 show the MAPLE, Minitab, and Excel plots for this polynomial respectively.

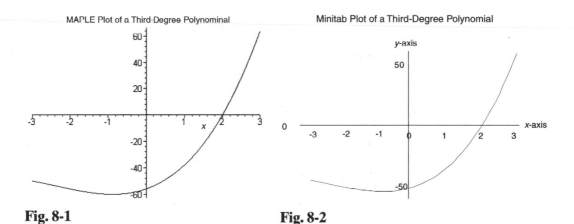

Fig. 8-1   **Fig. 8-2**

Note that the polynomial $y = x^3 + 7x^2 + 10x - 56$ intersects the x-axis at the point $x = 2$. This value is called a **root** of the polynomial. This confirms our finding in Example 8-6. If computer software is not available, a plot is made by computing the y values corresponding

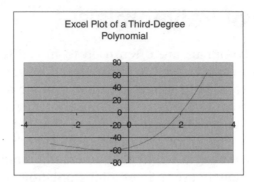

Fig. 8-3

to certain x values and then plotting those points in a rectangular coordinate system and connecting the points with a smooth curve. Table 8-6 gives some possible values that could be used to plot the polynomial $y = x^3 + 7x^2 + 10x - 56$.

Table 8-6   Some Points for Plotting $y = x^3 + 7x^2 + 10x - 56$

x	y	x	y
−3.0	−50.000	0.5	−49.125
−2.5	−52.875	1.0	−38.000
−2.0	−56.000	1.5	−21.875
−1.5	−58.625	2.0	0.000
−1.0	−60.000	2.5	28.375
−0.5	−59.375	3.0	64.000
0.0	−56.000		

Figure 8-4 shows the points from Table 8-6 plotted. If these points are connected, a graph very similar to those in Figs. 8-1, 8-2, and 8-3 is obtained.

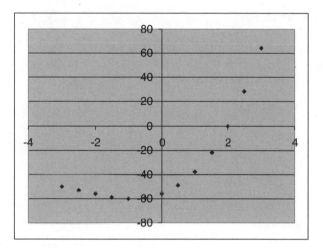

Fig. 8-4

Don't Forget

A *polynomial in one variable x* may be expressed as follows:

$$a_n x^n + a_{n-1} x^{n-1} + \cdots + a_1 x + a_0$$

The *variable* is x, the *degree* of the polynomial is n, the a's are numbers that are referred to as *coefficients*, and the product of a coefficient and a variable raised to a whole number power is called a *term* of the polynomial. The values of x that make the polynomial equal to zero are called *roots* of the polynomial.

To add two polynomials, combine like terms.

To subtract two polynomials, change the sign of each term in the polynomial being subtracted and then combine like terms.

To multiply two polynomials, multiply each term in one polynomial by each term in the other polynomial and then combine like terms.

To divide two polynomials, use the technique shown in the section Division of Polynomials.

Test Yourself

Questions

1. Consider the polynomial $y = 5x^4 - 3x^2 + 2x - 187$.

   (a) How many terms are there in the polynomial?

   (b) What is the degree of the polynomial?

2. Find $(x^3 - x^2 + 4x - 6) + (2x^3 + 3x^2 + x + 9)$.

3. Find $(x^3 - x^2 + 4x - 6) - (2x^3 + 3x^2 + x + 9)$.

4. Find $(x^3 - x^2 + 4x - 6)(2x^3 + 3x^2 + x + 9)$.

5. Find $x^3 + 4x^2 + x - 6$ divided by $x - 1$.

Answers

1. (a) Four;   (b) fourth

2. $3x^3 + 2x^2 + 5x + 3$

3. $-x^3 - 4x^2 + 3x - 15$

4. $2x^6 + x^5 + 6x^4 + 8x^3 - 23x^2 + 30x - 54$

5. $x^2 + 5x + 6$

CHAPTER 9

Quadratic Functions and Equations

Do I Need to Read This Chapter?

➡ What is a quadratic function?

➡ Where do the minimum and maximum values occur for a quadratic function?

➡ How do I use the quadratic formula to find the roots of a quadratic equation?

➡ How do I solve quadratic equations that have non-real-number solutions?

➡ What do the graphs of quadratic functions look like?

Quadratic Function

Example 9-1: MidwestUnivResearch.edu is a website that describes ongoing research projects at Midwest University. One such research project concerns the relationship between fertilizer application and corn yield. Table 9-1 gives the corn yield for differing amounts of fertilizer on comparable test plots.

Table 9-1　Corn Yield for Varying Amounts of Fertilizer Application

AMOUNT OF FERTILIZER (x)	CORN YIELD $y = -x^2 + 10x + 30$
0	30
1	39
2	46
3	51
4	54
5	55
6	54
7	51
8	46
9	39
10	30

A MAPLE plot of the data in Table 9-1 is shown in Fig. 9-1. The equation $y = -x^2 + 10x + 30$ is a second-degree polynomial. In algebra, a second-degree polynomial is called a **quadratic function.** The general form for a quadratic function is as follows, where a, b, and c are constants and a is not equal to zero.

$$y = ax^2 + bx + c$$

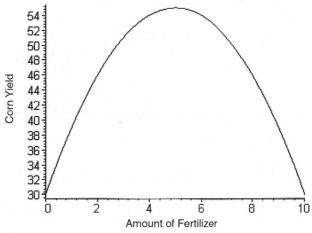

Fig. 9-1

Maximum Value of a Quadratic Function

Example 9-2: For the corn yield data, we have $a = -1$, $b = 10$, and $c = 30$. The graph of this quadratic function is called a **parabola.** If the value of a is negative, as in this example, the parabola will open downward. When five units of fertilizer was applied to a plot, the yield was 55 bushels of corn. This is the high point or **maximum yield** for the curve. This is referred to as the **vertex of the parabola.** In general, the maximum value of a quadratic function occurs when $x = -b/(2a)$. Note that in this case,

$$x = \frac{-b}{2a} = \frac{-(10)}{2(-1)} = 5$$

For values of x to the left of $x = 5$, the yield increases as the amount of fertilizer is increased. For values of x to the right of $x = 5$, the yield decreases as the amount of fertilizer is increased. It appears that there is an optimum yield that occurs when five units of fertilizer are applied and that when we go beyond that level of fertilizer, the yield begins to decrease.

Minimum Value for a Quadratic Function

Example 9-3: Toys-galore.com has one of the largest selections of toys available on the Internet. Table 9-2 gives the monthly sales in millions of dollars from December 1998 through December 1999.

Table 9-2   Sales for the Website Toys-galore.com

MONTH	MONTH, CODED x	SALES $y = x^2 - 13x + 55$
December 1998	1	43
January 1999	2	33
February 1999	3	25
March 1999	4	19
April 1999	5	15
May 1999	6	13
June 1999	7	13
July 1999	8	15
August 1999	9	19
September 1999	10	25
October 1999	11	33
November 1999	12	43
December 1999	13	55

Figure 9-2 gives a plot of the data shown in Table 9-2. The y-axis represents the monthly sales and the x-axis represents the month/year. The x-axis extends from December 1998 to December

1999. Such a plot is called a **time series.** If the months are coded as shown in Table 9-2, the equation $y = x^2 - 13x + 55$ may be used to describe the data. The plot for this quadratic function is shown in Fig. 9-3. Comparing this quadratic function with the general form, $y = ax^2 + bx + c,$ we see that $a = 1, b = -13,$ and $c = 55.$

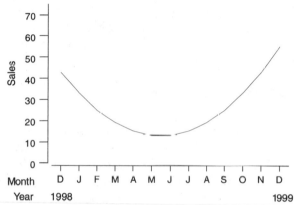

Fig. 9-2

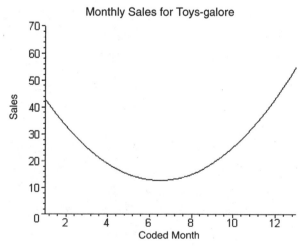

Fig. 9-3

Because a is positive, the parabola in Fig. 9-3 opens upward. The **minimum sales** occurs for $x = \dfrac{-b}{2a}$. For this quadratic function, $x = \dfrac{-(-13)}{2(1)} = 6.5$. The minimum sales is found by $y = (6.5)^2 - 13(6.5) + 55 = 12.75$. The vertex of the parabola is the point (6.5, 12.75). From Fig. 9-3 we see that to the left of the vertex sales decrease as x goes from 1 to 6.5. That is, sales are high during the Christmas season of 1998 and decrease as the summer season approaches. To the right of the vertex, sales increase as x goes from 6.5 to 13. That is, after the middle of summer sales begin to increase and become high again during the Christmas 1999 season.

Quadratic Equations

Example 9-4: Custom Boxes Inc. receives an order for 50 custom-made boxes with the following specifications: Each box is to be 5 inches in height, the length is to be 7 inches longer than the width, and the volume is to equal 150 cubic inches. What length and width would be needed to satisfy these specifications? If x represents the unknown width, then the length would be $x + 7$. The required specifications are shown in Fig. 9-4. The volume requirement leads to the equation

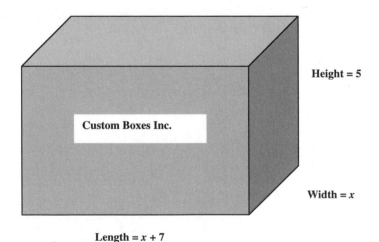

Height = 5

Custom Boxes Inc.

Width = x

Length = $x + 7$

Fig. 9-4

$$5x(x + 7) = 150$$

since the volume $V = lwh$ is required to equal 150. This equation may be expressed as

$$5x^2 + 35x - 150 = 0$$

Such an equation is called a **quadratic equation.**

Pattern

The general form of a quadratic equation is

$$ax^2 + bx + c = 0$$

Such an equation has two solutions, called the *roots* of the equation, given by the *quadratic formula* as follows.

$$x = \frac{-b \pm \sqrt{b^2 - 4ac}}{2a}$$

For the equation $5x^2 + 35x - 150 = 0$, $a = 5$, $b = 35$, and $c = -150$. Substituting into the quadratic formula we obtain

$$x = \frac{-35 \pm \sqrt{35^2 - 4(5)(-150)}}{2(5)}$$

Now, the quantity under the square root is $35^2 - 4(5)(-150) = 4225$. Using a calculator, we find that the square root of 4225 is 65. Substituting into the quadratic formula, we find the following:

$$x = \frac{-35 \pm 65}{10}$$

Using the plus sign we find $x = \dfrac{30}{10} = 3$ and using the minus sign we find $x = \dfrac{-100}{10} = -10$. Since the width can not be negative, we select $x = 3$ as our solution. To see that our solution is correct, we note that if the width is 3, the length is 10, and the height is 5, the volume is $V = 10(3)(5) = 150$.

Short Cuts

MAPLE software utilizes the quadratic formula to solve the quadratic equation. The MAPLE solution is as follows.

```
> solve( 5*x^2 + 35*x - 150=0 );
-10, 3
```

Quadratic Equations with Non-Real Number Solutions

Danger!

Some quadratic equations do not have any real numbers that are solutions to the equation. In attempting to find solutions to such equations, mathematicians introduce a whole new system of numbers called *complex numbers.* At first sight, such numbers seem removed from any real-world applications. That is, they do not appear be useful in solving problems. However, it turns out that such numbers are extremely useful in solving problems in disciplines of study such as electrical engineering. In fact, mathematics majors often take a course called Complex Variables dealing with such numbers.

Example 9-5: There are no real numbers that satisfy the quadratic equation

$$x^2 + 1 = 0 \quad \text{or} \quad x^2 = -1$$

No matter how hard you try, you cannot find a real number that makes this equation true, because a nonzero real number squared will always be positive. However, if we apply the quadratic formula to this equation with $a = 1$, $b = 0$, and $c = 1$, we find the following:

$$x = \frac{-b \pm \sqrt{b^2 - 4ac}}{2a} = \frac{-0 \pm \sqrt{0^2 - 4(1)(1)}}{2(1)}$$

$$x = \pm \frac{\sqrt{-4}}{2}$$

There is no real number whose square is –4!

Quick Tip

Here is how the mathematicians proceeded to solve this problem. The square root of –4 is expressed as a product of square roots as follows.

$$\sqrt{-4} = \sqrt{-1}\sqrt{4}$$

The square root of –1 is given the name *imaginary unit* and is represented by the letter *i*.

$$i = \sqrt{-1}$$

With this imaginary unit we express the square root of –4 as follows:

$$\sqrt{-4} = \sqrt{-1}\sqrt{4} = 2i$$

and the solution to the equation $x^2 + 1 = 0$ is

$$x = \pm \frac{\sqrt{-4}}{2} = \pm \frac{2i}{2} = \pm i$$

If both sides of the equation defining *i* is squared, the following is obtained

$$i^2 = -1$$

We see that $x = i$ is a solution to the equation $x^2 + 1 = 0$, since

$$i^2 + 1 = -1 + 1 = 0$$

Example 9-6: To investigate further the idea of quadratic equations with non-real number solutions, consider the quadratic equation

$$x^2 - 4x + 5 = 0$$

If we apply the quadratic formula to this equation with $a = 1$, $b = -4$, and $c = 5$, we find the following.

$$x = \frac{-b \pm \sqrt{b^2 - 4ac}}{2a} = \frac{-(-4) \pm \sqrt{(-4)^2 - 4(1)(5)}}{2(1)}$$

Simplifying, we find

$$x = \frac{4 \pm \sqrt{-4}}{2} = \frac{4 \pm 2\mathbf{i}}{2}$$

The two solutions are $2 + \mathbf{i}$ and $2 - \mathbf{i}$, since

$$\frac{4 \pm 2\mathbf{i}}{2} = \frac{2(2 \pm \mathbf{i})}{2} = 2 \pm \mathbf{i}$$

Note that these numbers are of the form $a + b\mathbf{i}$, where a and b are real numbers and $\mathbf{i}$ is the imaginary unit. Numbers of this form are called **complex numbers.** The next chapter will study complex numbers in more detail.

Graphs of Quadratic Functions

This section will illustrate the graphical implications of quadratic functions. The vertex and roots will be noted and the use of software to produce graphs of quadratic functions will be considered. Figure 9-5 is a MAPLE plot of the quadratic function $y = x^2 + x - 6$. Applying the quadratic formula with $a = 1$, $b = 1$, and $c = -6$ to the quadratic equation

$$x^2 + x - 6 = 0$$

we find the roots to be $x = -3$ and $x = 2$. Note that this is where the parabola cuts the x-axis and therefore $y = 0$ for these values. The fact that a is positive tells us that the parabola will open upward. The x value of the vertex is given by $x = \dfrac{-b}{2a}$ or $x = -1/2$. The minimum value for y occurs at this value of x and is equal to

$$y = \left(-\frac{1}{2}\right)^2 + \left(-\frac{1}{2}\right) - 6 = -6.25$$

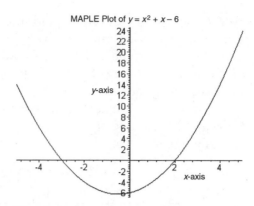

Fig. 9-5

Figure 9-6 is a MINITAB plot of the quadratic function $y = -x^2 + 10x$. Applying the quadratic formula with $a = -1$, $b = 10$, and $c = 0$ to the quadratic equation

$$-x^2 + 10x = 0$$

we find the roots to be $x = 0$ and $x = 10$. Note that this is where the parabola cuts the x-axis and therefore $y = 0$ for these values. The fact that a is negative tells us that the parabola will open downward. The x value of the vertex is given by $x = \dfrac{-b}{2a}$ or $x = 5$. The maximum value for y occurs at this value of x and is equal to

$$y = -(5)^2 + 10(5) = 25$$

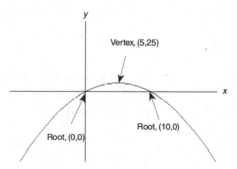

Fig. 9-6

Figure 9-7 is an Excel plot of the quadratic function $y = x^2 - 4x + 25$. Applying the quadratic formula with $a = 1$, $b = -4$, and $c = 25$, we find that the roots are complex. They are found to be $2 - \sqrt{21}\,\mathbf{i}$ and $2 + \sqrt{21}\,\mathbf{i}$. From Fig. 9-7 we see that the parabola does not cross the x-axis. This corresponds to the fact that the quadratic equation has no real roots. The fact that a is positive tells us that the parabola will open upward. The x value of the vertex is $x = -b/2a$ or in this case $x = 2$. The minimum value for y occurs at this value of x and is equal to

$$y = (2)^2 - 4(2) + 25 = 21$$

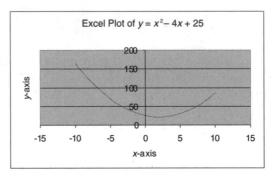

Fig. 9-7

Don't Forget

The general form of a *quadratic function* is $y = ax^2 + bx + c$. The graph of a quadratic function is called a *parabola*. If a is positive, the parabola opens upward and the *minimum value* for y occurs at the vertex of the parabola. The x value of the vertex is given by $x = -b/(2a)$. If a is negative, the parabola opens downward and the *maximum value* for y occurs at the vertex of the parabola.

A *quadratic equation* is of the form $ax^2 + bx + c = 0$. The two values of x for which this equation is true are called the *roots* of the quadratic equation. They are given by the quadratic formula as follows:

$$x = \frac{-b \pm \sqrt{b^2 - 4ac}}{2a}$$

Pattern

The quantity $b^2 - 4ac$ is called the *discriminant* for the quadratic equation. If the discriminant is negative, the roots of the quadratic equation will contain the imaginary unit. If the discriminant is zero, the same real root occurs twice. If the discriminant is positive, there are two unequal real roots.

The *imaginary unit i* is defined to equal the square root of -1. Numbers of the form $a + bi$, where a and b are real numbers, are called *complex numbers*.

When a quadratic function is graphed, the resulting parabola will have either a minimum or a maximum value that occurs at the vertex. If the parabola intersects the x-axis, these will be the real roots of a quadratic equation.

Test Yourself

Questions

Answer Questions 1 through 4 for both quadratic function 1 and quadratic function 2. Quadratic function 1: $y = 3x^2 + 6x - 24$. Quadratic function 2: $y = -2x^2 + 4x + 16$.

1. Does the graph of the quadratic function open upward or downward?

2. What are the coordinates of the vertex?

3. What is maximum/minimum y value for the function?

4. Where does the graph of the function intercept the x-axis?

5. Find the roots to the quadratic equation $x^2 + 4 = 0$.

Answers

1. Quadratic function 1:
   upward

   Quadratic function 2:
   downward

2. Quadratic function 1:
   $(-1, -27)$

   Quadratic function 2:
   $(1, 18)$

3. Quadratic function 1:   Quadratic function 2:
   minimum = −27.   maximum = 18.

4. Quadratic function 1:   Quadratic function 2:
   $x = -4$ and $x = 2$. $x = -2$ and $x = 4$.

5. −2**i** and 2**i**

CHAPTER 10

Roots, Radicals, and Complex Numbers

Do I Need to Read This Chapter?

➡️ What are square roots, cube roots, and higher-order roots?

➡️ What do I know about the sum, difference, product, and quotient of radicals?

➡️ What are complex numbers and how do I add, subtract, multiply, and divide them?

➡️ What do MAPLE and Excel plots of polynomials with both real and imaginary roots look like?

Square Roots

Example 10-1: CustomRugs.com is an Internet location specializing in custom-designed rugs. Suppose a square rug of area 121 square feet is ordered. Because the rug is square, the dimension of the rug may be found by finding the square root of the area. The square root is indicated by either the square root radical, $\sqrt[2]{}$, or by the exponent 1/2. The 2 is often omitted in the square root radical and the symbol $\sqrt{}$ is used. In general, b is the square root of a if $b^2 = a$. That is the following two statements are equivalent.

$$\sqrt{a} = b \quad \text{is equivalent to} \quad b^2 = a$$

In the case of the rug having area 121 square feet, there are two real numbers that represent the square root of 121, since

$$(11)^2 = 121 \quad \text{and} \quad (-11)^2 = 121$$

The positive square root of 121 is 11. Because measurements of dimensions are positive, the answer 11 would be chosen as the answer and the rug is an 11×11 rug. The positive square root of a number is called the **principal square root.**

Cube Roots

Example 10-2: Custom Boxes Inc. receives orders for cubical boxes of various volumes from different companies. Table 10-1

Table 10-1 Some of the Boxes Available from CustomBoxes.com

VOLUME	DIMENSIONS	LENGTH OF SIDE
8	2 by 2 by 2	2
27	3 by 3 by 3	3
64	4 by 4 by 4	4
125	5 by 5 by 5	5
216	6 by 6 by 6	6

gives the dimensions of various boxes with volumes equal to 8, 27, 64, 125, and 216 cubic inches.

This table may be used to determine the dimensions of a box that has a particular volume. Note that each of these volumes is a perfect cube. The volume is equal to the length of a side cubed. Conversely, the length of a side is equal to the cube root of the volume. The cube root of the volume is written as $\sqrt[3]{\text{volume}}$ or as the volume to the 1/3 power. That is, $(\text{volume})^{1/3}$ represents the cube root of the volume. The symbol $\sqrt[3]{}$ is called **radical notation** for cube root. In general, b is the cube root of a if $b^3 = a$. That is the following two statements are equivalent:

$$\sqrt[3]{a} = b \qquad \text{is equivalent to} \qquad b^3 = a$$

What dimensions are needed for a box having a volume equal to 100 cubic inches? We need to find the length of a side, represented by x, that is such that

$$x^3 = 100 \qquad \text{or} \qquad x = \sqrt[3]{100}$$

Since $4^3 = 64$ and $5^3 = 125$, we see that x must be between the real numbers 4 and 5. Because there is no whole number whose cube equals 100, we must use a calculator or computer software to find the answer.

Pattern

The function $\sqrt[x]{y}$ found on most calculators is used as follows to find the cube root of 100. The following steps will result in finding the cube root of 100.

Step 1: Enter 100, the number for which you wish to find the cube root.
Step 2: Select the function $\sqrt[x]{y}$.
Step 3: Enter 3 for the cube root.
Step 4: Enter = . The result is 4.64 to two decimal places.

Multiplying 4.64 times itself three times gives the result 99.897344. The more decimal places you use, the closer will be the cube of the number to 100.

Short Cuts

A more complete picture of the relationship between the volumes required for the cubical boxes ordered by various companies and the length of a side for the cubical box is provided by a the graph of the cube root function. The cube root function is as follows:

$$y = \sqrt[3]{x} = x^{1/3}$$

The following MAPLE command produces the plot shown in Fig. 10-1.

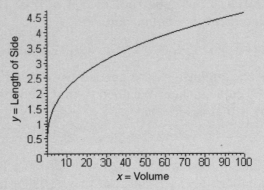

Fig. 10-1

```
➤ plot( x^(1/3), x=0..100,tickmarks=[15,10],labels =
['x = volume','y = length of
side'] );
```

See if you can locate the points given in Table 10-1 on the graph in Fig. 10-1.

Higher-Order Roots

The symbol $\sqrt[n]{}$ is called **radical notation** for the nth root. In general, b is the nth root of a if $b^n = a$. That is, the following two statements are equivalent:

$$\sqrt[n]{a} = b \qquad \text{is equivalent to} \qquad b^n = a$$

Example 10-3: If principal P is invested at interest rate $r\%$ and the interest is compounded yearly for n years, the value of the investment after n years is given by the following equation:

$$\text{Value of investment after } n \text{ years} = P(1 + r/100)^n$$

Suppose \$5000 is invested in a mutual fund and nothing is added to or taken from the fund for 5 years. If the value of the fund at the end of 5 years is \$10,000, what equivalent yearly rate of compound interest did the fund yield? We are asked to find r. The equation of value is

$$10,000 = 5000(1 + r/100)^5$$

Dividing both sides by 5000 we obtain:

$$2 = (1 + r/100)^5$$

To solve this equation we need to take the fifth root of both sides. This gives

$$\sqrt[5]{2} = 1 + r/100$$

Using a calculator and the procedure for finding the fifth root, we find that $\sqrt[5]{2} = 1.148698$. The last equation now becomes

$$1.148698 = 1 + \frac{r}{100}$$

or solving for r, we find

$$r = 14.9\%$$

Danger!

There are no real even roots of negative numbers. That is if a is less than zero and n is an even whole number, there is no real number b that satisfies $b^n = a$. Suppose you wished to find the fourth root of -16. Even though $2^4 = 16$ and $(-2)^4 = 16$, there is no real number b that satisfies the equation $b^4 = -16$.

Four Fundamental Operations Involving Radicals

The nth root of a product of two nonnegative quantities is equal to the product of the nth roots of the quantities. This is expressed as follows.

$$\sqrt[n]{ab} = \sqrt[n]{a}\sqrt[n]{b} \quad \text{for} \quad a \geq 0 \quad \text{and} \quad b \geq 0$$

A similar result holds for quotients and is as follows:

$$\sqrt[n]{\frac{a}{b}} = \frac{\sqrt[n]{a}}{\sqrt[n]{b}} \quad \text{for} \quad a \geq 0 \quad \text{and} \quad b > 0$$

Example 10-4: The following illustrates these laws for square roots:

$$\sqrt{4(25)} = \sqrt{100} = 10$$
$$\sqrt{4}\sqrt{25} = 2(5) = 10$$
$$\sqrt{\frac{100}{25}} = \sqrt{4} = 2$$
$$\frac{\sqrt{100}}{\sqrt{25}} = \frac{10}{5} = 2$$

Danger!

The nth root of a sum or difference is not equal to the sum or difference of the nth roots.

Example 10-5: This example illustrates this potential trouble spot.

$$\sqrt{16+9} = \sqrt{25} = 5$$
$$\sqrt{16} + \sqrt{9} = 4 + 3 = 7$$
$$\sqrt{16+9} \neq \sqrt{16} + \sqrt{9}$$

Complex Numbers

In our discussion of the solutions of quadratic equations, we found that some solutions involved the imaginary unit **i**. In that discussion, we found that **i** is defined as follows.

$$i = \sqrt{-1} \quad \text{and} \quad i^2 = -1$$

Additional powers of **i** provide some interesting results.

$$i^3 = i^2(i) = (-1)(i) = -i$$
$$i^4 = i^2(i^2) = (-1)(-1) = 1$$

If this process is continued, it is found that any power of **i** is either **i**, **−i**, 1 or −1. The definition of a complex number is any number of the form $a + b\mathbf{i}$, where a and b are real numbers. The number a is called the **real part of the complex number** and the number b is called the **imaginary part of the complex number.** Complex numbers have applications in many fields of science and engineering.

Operations on Complex Numbers

Example 10-6: To add or subtract two complex numbers, you add or subtract the real part and add or subtract the imaginary parts. The following examples illustrate how to add and subtract complex numbers.

$$(3 + 2\mathbf{i}) + (5 + 7\mathbf{i}) = (8 + 9\mathbf{i})$$
$$(3 + 2\mathbf{i}) - (5 + 7\mathbf{i}) = (-2 - 5\mathbf{i})$$

We multiply two complex numbers using the FOIL method discussed in Chapter 1 and then replace i^2 by −1. To multiply $(3 - \mathbf{i})$ times $(5 + 2\mathbf{i})$, we proceed as follows.

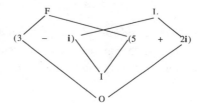

Summarizing, we have the following:

$$(3-\mathbf{i})(5+2\mathbf{i}) = 15 + 6\mathbf{i} - 5\mathbf{i} - 2\mathbf{i}^2$$
$$(3-\mathbf{i})(5+2\mathbf{i}) = 15 + \mathbf{i} - 2(-1)$$
$$(3-\mathbf{i})(5+2\mathbf{i}) = 17 + \mathbf{i}$$

In order to describe the division of two complex numbers, we need to define the **conjugate of a complex number.** The conjugate of a complex number is obtained by changing the sign of the imaginary part of the complex number. To divide two complex numbers, multiply the numerator and denominator by the conjugate of the denominator. Suppose we wish to divide $3 - \mathbf{i}$ by $5 + \mathbf{i}$. The conjugate of $5 + \mathbf{i}$ is $5 - \mathbf{i}$. The division is accomplished as follows:

$$\frac{3-\mathbf{i}}{5+\mathbf{i}} = \frac{3-\mathbf{i}}{5+\mathbf{i}} \cdot \frac{5-\mathbf{i}}{5-\mathbf{i}}$$

Multiplying the numerators and denominators, we obtain:

$$\frac{3-\mathbf{i}}{5+\mathbf{i}} \cdot \frac{5-\mathbf{i}}{5-\mathbf{i}} = \frac{15 - 5\mathbf{i} - 3\mathbf{i} + \mathbf{i}^2}{25 + 5\mathbf{i} - 5\mathbf{i} - \mathbf{i}^2} = \frac{14 - 8\mathbf{i}}{26} = \frac{14}{26} - \frac{8}{26}\mathbf{i} = \frac{7}{13} - \frac{4}{13}\mathbf{i}$$

Note again that any time $\mathbf{i}^2$ appears, it is replaced by -1.

Quick Tip

It is difficult for algebra students to see the physical significance (real-world applications) of operations on complex numbers. Unfortunately, the useful applications of complex numbers occurs in more advanced science and engineering courses.

Short Cuts

Example 10-7: Let us consider once again the power of the MAPLE software to perform algebraic operations by using MAPLE to perform the operations on complex numbers illustrated in Example 10-6.

```
>  (3+2*I)+(5+7*I);            8 + 9 I
>  (3+2*I)-(5+7*I);            -2 - 5 I
>  (3-I)*(5+2*I);              17 + I
>  (3-I)/(5+I);                7/13 - 4/13 I
```

Polynomials with Both Real and Imaginary Roots

Example 10-8: Consider the following fourth-degree polynomial.

$$y = -x^4 + x^3 + x^2 + x + 2$$

Table 10-2 gives some points on this polynomial that may be used to form a plot of the polynomial if computer software is not available.

Table 10-2   Points on $y = -x^4 + x^3 + x^2 + x + 2$

x	y	x	y	x	y	x	y	x	y
−2	−20	−1	0	0	2	1	4	2	0
−1.75	−11	−0.75	1.1	0.25	2.3	1.25	4.3	2.25	−4.9
−1.5	−5.7	−0.5	1.6	0.5	2.8	1.5	4.1	2.5	−13
−1.25	−2.1	−0.25	1.8	0.75	3.4	1.75	2.8	2.75	−24

Figure 10-2 show an Excel plot of these points. If they are now connected, a graph similar to the MAPLE plot of $y = -x^4 + x^3 + x^2 + x + 2$ shown in Fig. 10-3 is obtained.

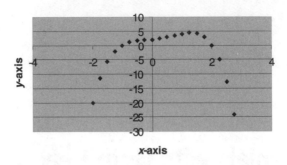

Fig. 10-2

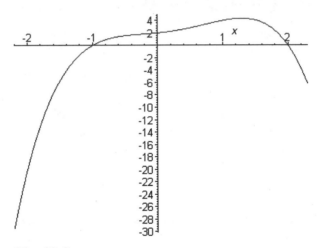

Fig. 10-3

The roots (values of x that make the polynomial equal to zero) are found using MAPLE and the output is as follows.

```
> solve( x^4-x^3-x^2-x-2=0 );   2, -1, I, -I
```

Thus the roots are $2, -1,$ **i,** and **–i.**

It is instructional to verify that these are indeed the roots of the polynomial. For the real root $x = -1$, we have

$$P(-1) = -(-1)^4 + (-1)^3 + (-1)^2 + (-1) + 2 = -1 - 1 + 1 - 1 + 2 = 0$$

For the real root $x = 2$, we have

$$P(2) = -(2)^4 + (2)^3 + (2)^2 + (2) + 2 = -16 + 8 + 4 + 2 + 2 = 0$$

For the imaginary root $x = \mathbf{i}$, we have

$$P(\mathbf{i}) = -(\mathbf{i})^4 + (\mathbf{i})^3 + (\mathbf{i})^2 + (\mathbf{i}) + 2 = -(1) + (-\mathbf{i}) + (-1) + (\mathbf{i}) + 2 = 0$$

Similarly, for the imaginary root $x = -\mathbf{i}$, $P(-\mathbf{i}) = 0$.

Note that the real roots are the points where the polynomial cuts the x-axis. However, the imaginary roots do not have a similar interpretation.

Don't Forget

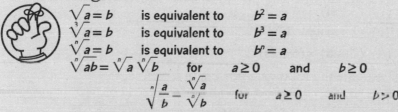

$\sqrt{a} = b$ is equivalent to $b^2 = a$

$\sqrt[3]{a} = b$ is equivalent to $b^3 = a$

$\sqrt[n]{a} = b$ is equivalent to $b^n = a$

$\sqrt[n]{ab} = \sqrt[n]{a}\,\sqrt[n]{b}$ for $a \geq 0$ and $b \geq 0$

$\sqrt[n]{\dfrac{a}{b}} = \dfrac{\sqrt[n]{a}}{\sqrt[n]{b}}$ for $a \geq 0$ and $b > 0$

The nth root of a sum or difference is not equal to the sum or difference of the nth roots.

The definition of a *complex number* is any number of the form $a + bi$, where a and b are real numbers. The number a is called the *real part* of the complex number and the number b is called the *imaginary part* of the complex number.

To add or subtract two complex numbers, you add or subtract the real parts and add or subtract the imaginary parts.

Complex numbers are multiplied in the same manner as polynomials are multiplied.

Complex numbers are divided by multiplying the numerator and denominator by the *conjugate* of the denominator.

Test Yourself

Questions

1. Find the principal square roots of the following numbers:
   (a) 25; (b) 81; (c) 35

2. Find the cube roots of the following numbers:
   (a) 64; (b) 125; (c) 35

3. Compute the values of the two quantities given in the following:
   (a) $\sqrt[3]{64(125)}$ and $\sqrt[3]{64}(\sqrt[3]{125})$

   (b) $\sqrt[3]{\dfrac{125}{64}}$ and $\dfrac{\sqrt[3]{125}}{\sqrt[3]{64}}$

   (c) $\sqrt[3]{125 + 64}$ and $\sqrt[3]{125} + \sqrt[3]{64}$

4. Perform the following operations on the complex numbers:
   (a) $(2 + 4i) + (6 - 2i)$
   (b) $(2 + 4i) - (6 - 2i)$
   (c) $(2 + 4i)(6 - 2i)$
   (d) $(2 + 4i) \div (6 - 2i)$

5. Evaluate the following powers of **i**:
   (a) i^{35}
   (b) i^{36}
   (c) i^{37}
   (d) i^{38}

Answers

1. (a) 5;   (b) 9;   (c) 5.916079 . . .
2. (a) 4;   (b) 5;   (c) 3.271066 . . .
3. (a) 20 and 20;   (b) 1.25 and 1.25;   (c) 5.738794 . . . and 9
4. (a) $8 + 2i$;   (b) $-4 + 6i$;   (c) $20 + 20i$;   (d) $0.1 + 0.7i$
5. (a) $-i$;   (b) 1;   (c) **i**;   (d) -1

CHAPTER 11

Rational Expressions

Do I Need to Read This Chapter?

➜ What is a rational expression?

➜ How do I multiply, divide, add, and subtract rational expressions?

➜ How do I solve equations containing rational expressions?

Rational Expressions

Example 11-1: Booksales.com is a large seller of books. The marketing group at Booksales.com knows that if a book is priced low the demand will be high, but the profit per book will be low. On the other hand, if the book is priced high, the demand will be low and the profit per book will be high. Table 11-1 gives several different possible selling prices, the sales demand in thousands, the profit in thousands, and the average profit per book for a new paperback.

Table 11-1

PRICE PER BOOK p	DEMAND (THOUSANDS) $16 - p$	PROFIT (THOUSANDS) $-p^2 + 25p - 144$	AVERAGE PROFIT PER BOOK $p - 9$
10.00	6	6	1
11.00	5	10	2
12.00	4	12	3
13.00	3	12	4
14.00	2	10	5
15.00	1	6	6

The first row of the table indicates that if the book is sold for $10.00, the demand will be 6000, the profits will be $6000, and the average profit per book will be $1.00. In general, if the price per book is p, the demand is $16 - p$, the profit is $-p^2 + 25p - 144$. The average profit per book is equal to the profit divided by the demand. Figure 11-1 is a Minitab plot showing the relationship among price, demand, profit, and average profit per book.

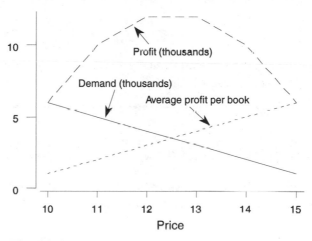

Fig. 11-1

The average profit per book is given by the following rational expression:

$$\frac{-p^2 + 25p - 144}{16 - p}$$

A **rational expression** is defined to be the quotient of two polynomials. The above rational expression is undefined for $p = 16$. Such a value for which the denominator of the rational expression equals zero is called an **excluded value** for the expression. The numerator of this rational expression is factored as follows:

$$-p^2 + 25p - 144 = (p - 9)(16 - p)$$

Substituting in the above rational expression, we obtain the following simplified form for the rational expression. The rational expression is reduced to lowest terms.

$$\frac{(p - 9)(16 - p)}{16 - p} = p - 9$$

Danger!

When reducing a rational expression to lowest terms, only factors of both the numerator and denominator may be divided out (canceled).

$$\frac{\cancel{x}(x+2)(x-3)}{\cancel{x}+(x+10)} \neq \frac{(x+2)(x-3)}{(x+10)}$$

The x in the numerator and the x in the denominator may not be divided out because the x in the denominator is not a factor.

Multiplying Rational Expressions

Example 11-2: Suppose the demand (in thousands) for a compact disc is given by the rational expression

$$\frac{30}{x+6}$$

where x is the price per compact disc and the average profit per compact disc on sales of the compact disc is given by

$$\frac{x^2+5x-6}{5x+10}$$

Then the total profit (in thousands) is equal to the demand times the average profit per unit and is given by the following product of rational expressions:

$$\left(\frac{30}{x+6}\right)\left(\frac{x^2+5x-6}{5x+10}\right)=\left(\frac{30}{x+6}\right)\left(\frac{(x+6)(x-1)}{5(x+2)}\right)=\frac{6(x-1)}{x+2}$$

The factor $x + 6$ is divided out in both the numerator and the denominator and the 5 is divided into the 30 to obtain the 6. Table 11-2 gives the demand, average profit, and total profit for various selling prices. Figure 11-2 shows a plot of demand, total profit, and average profit versus price for Example 11-2.

Table 11-2

PRICE	DEMAND (THOUSANDS)	AVERAGE PROFIT	TOTAL PROFIT (THOUSANDS)
x	$\dfrac{30}{x+6}$	$\dfrac{x^2+5x-6}{5x+10}$	$\dfrac{6(x-1)}{x+2}$
5.0	2.7	1.3	3.4
5.5	2.6	1.4	3.6
6.0	2.5	1.5	3.8
6.5	2.4	1.6	3.9
7.0	2.3	1.7	4.0
7.5	2.2	1.8	4.1
8.0	2.1	2.0	4.2
8.5	2.1	2.1	4.3
9.0	2.0	2.2	4.4
9.5	1.9	2.3	4.4
10.0	1.9	2.4	4.5

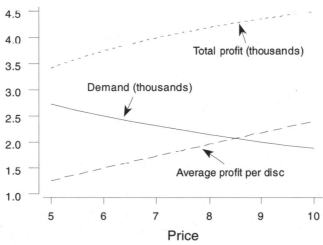

Demand, Total Profit, Average Profit as a Function of Price

Fig. 11-2

Pattern

Two rational expressions are multiplied just as two fractions are multiplied. That is, we multiply the numerators and the denominators, and then reduce the result to lowest terms.

Dividing Rational Expressions

Example 11-3: Suppose the total profit on sales of a software program is given by $\dfrac{6x-6}{x+2}$, where x is the price of the software program. Suppose that when x is the price, the demand is given by $\dfrac{30}{x+6}$. The average profit is given by the total profit divided by the demand as follows:

$$\frac{6x-6}{x+2} \div \frac{30}{x+6}$$

The division of two rational expressions is performed just as the division of two fractions. That is, invert the second fraction and then multiply.

$$\left(\frac{6x-6}{x+2}\right)\left(\frac{x+6}{30}\right) = \frac{6(x-1)(x+6)}{30(x+2)}$$

$$\frac{\cancel{6}(x-1)(x+6)}{\cancel{30}(x+2)} = \frac{(x-1)(x+6)}{5(x+2)}$$

$$\frac{(x-1)(x+6)}{5(x+2)} = \frac{x^2+5x-6}{5x+10}$$

Pattern

To divide two rational fractions, invert the second and then multiply. After multiplying, reduce to lowest terms by dividing out common factors in the numerator and denominator.

Addition and Subtraction of Rational Expressions

In order to add or subtract fractions, we first obtain a lowest common denominator (LCD) and then combine numerators over that LCD. For example, suppose we wish to find the following:

$$\frac{3}{14} + \frac{2}{15} - \frac{4}{18}$$

The LCD is found by considering the prime factorization for each denominator.

$$14 = 2(7) \qquad 15 = 3(5) \qquad \text{and} \qquad 18 = 2(3)(3)$$

The LCD is the product of the prime factors as follows:

$$\text{LCD} = 2(3)(3)(5)(7) = 630$$

The fractions are converted to equivalent fractions having a denominator equal to 630 as follows:

$$\frac{3}{14} + \frac{2}{15} - \frac{4}{18} = \frac{135}{630} + \frac{84}{630} - \frac{140}{630} = \frac{79}{630}$$

The same technique is used to add or subtract rational expressions.

Example 11-4: Combine the following rational expressions:

$$\frac{6}{x-1} - \frac{4}{x-2} + \frac{4}{x^2 - 3x + 2}$$

The factorization for $x^2 - 3x + 2$ is $(x-1)(x-2)$. The factors to occur in the denominators are $(x-1)$ and $(x-2)$ and the LCD $= (x-1)(x-2)$. The original problem may be expressed as follows, where each fraction is expressed with the same LCD.

$$\frac{6}{x-1} - \frac{4}{x-2} + \frac{4}{x^2 - 3x + 2}$$

$$= \frac{6(x-2)}{(x-1)(x-2)} - \frac{4(x-1)}{(x-1)(x-2)} + \frac{4}{(x-1)(x-2)}$$

Combining numerators over the LCD, we have

$$\frac{6(x-2)}{(x-1)(x-2)} - \frac{4(x-1)}{(x-1)(x-2)} + \frac{4}{(x-1)(x-2)} = \frac{6(x-2) - 4(x-1) + 4}{(x-1)(x-2)}$$

$$\frac{6(x-2) - 4(x-1) + 4}{(x-1)(x-2)} = \frac{6x - 12 - 4x + 4 + 4}{(x-1)(x-2)}$$

$$\frac{6x - 12 - 4x + 4 + 4}{(x-1)(x-2)} = \frac{2x - 4}{(x-1)(x-2)}$$

$$\frac{2x - 4}{(x-1)(x-2)} = \frac{2(x-2)}{(x-1)(x-2)} = \frac{2}{(x-1)}$$

Finally, we have the following result:

$$\frac{6}{x-1} - \frac{4}{x-2} + \frac{4}{x^2 - 3x + 2} = \frac{2}{(x-1)}$$

This equality holds for every value of x except the excluded values $x = 1$ and $x = 2$. Suppose $x = 3$ for instance. The value for the left-hand side of the equation is

$$\frac{6}{3-1} - \frac{4}{3-2} + \frac{4}{3^2 - 3(3) + 2} = 3 - 4 + 2 = 1$$

The value for the right-hand side is $\dfrac{2}{(3-1)}$ which also equals 1.

Pattern

To add or subtract rational expressions, use the following steps.

1. Find the LCD for all the rational expressions.
2. Change all given rational expressions to equivalent rational expressions, with the LCD as their denominators.
3. Add or subtract the rational expressions.
4. Reduce the result to lowest terms.

Short Cuts

The solution to Example 11-4 using MAPLE is as follows:

```
> simplify( 6/(x-1) - 4/(x-2) + 4/(x^2-3*x+2) );
        2
      -----
      x - 1
```

Solving Equations Containing Rational Expressions

Example 11-5: In Example 11-1, the average profit per book as a function of the price per book was given by

$$\frac{-p^2 + 25p - 144}{16 - p}$$

Suppose we wish to determine the price we need to charge in order that the average profit equal \$1.50 per book. This requirement leads to the equation:

$$\frac{-p^2 + 25p - 144}{16 - p} = 1.50$$

To solve this equation, first multiply both sides of the equation by $16 - p$ to clear fractions. The following equation is obtained

$$-p^2 + 25p - 144 = (16 - p)(1.50) = 24 - 1.5p$$

Rearranging terms and writing in the form of a quadratic equation, we obtain:

$$-p^2 + 26.5p - 168 = 0$$

This equation may be solved by using the quadratic equation or by using MAPLE. The solution using MAPLE is:

```
> solve( -p^2+26.5*p-168=0);
              10.50000000, 16.
```

Because the expression $\dfrac{-p^2 + 25p - 144}{16 - p}$ is undefined when $p = 16$, the solution is $p = \$10.50$. This solution could also be obtained by noting that the expression $\dfrac{-p^2 + 25p - 144}{16 - p}$ reduces to $p - 9$. And solving $p - 9 = 1.50$ for p also gives $p = \$10.50$.

Pattern

To solve an equation containing rational expressions use the following steps.

1. Clear all rational expressions by multiplying each term on both sides of the equation by the LCD of the rational expressions.

2. Solve the resulting equation.

3. Check all answers to see if they satisfy the original equation.

Don't Forget

A *rational expression* is a ratio of two polynomials.

An *excluded value* for a rational expression is a value for which the denominator of the rational expression is equal to zero.

When performing the four basic operations of addition, subtraction, multiplication, and division on rational expressions, the same rules are followed as when performing the four basic operations on numerical fractions.

To multiply two rational expressions, multiply the numerators and the denominators and then reduce to lowest terms.

To divide two rational expressions, invert the second and then multiply.

To add or subtract rational expressions, use the following steps.
1. Find the LCD for all the rational expressions.
2. Change all given rational expressions to equivalent rational expressions, with the LCD as their denominators.
3. Add or subtract the rational expressions.
4. Reduce the result to lowest terms.

To solve an equation containing rational expressions use the following steps:
1. Clear all rational expressions by multiplying each term on both sides of the equation by the LCD of the rational expressions.
2. Solve the resulting equation.
3. Check all answers to see if they satisfy the original equation.

Test Yourself

Questions

1. Verify the following factorizations of the numerator and the denominator of the rational expression and reduce the rational expression to lowest terms. Identify all excluded values.

$$\frac{x^2 + 3x + 2}{x^3 - 7x - 6} = \frac{(x+1)(x+2)}{(x+1)(x+2)(x-3)}$$

2. Verify the following factorizations of the rational expressions and then multiply the two rational expressions.

$$\frac{x^2 - 2x - 63}{x^2 - 4x - 45} = \frac{(x-9)(x+7)}{(x-9)(x+5)}$$

$$\frac{2x^2 + 13x + 15}{2x^2 - 5x - 12} = \frac{(2x+3)(x+5)}{(2x+3)(x-4)}$$

3. Perform the following division:
$$\frac{x^2 - 1}{x^2 + 4x + 3} \div \frac{x^2 - 2x - 3}{x^2 - 9}$$

4. Perform the following addition:
$$\frac{1}{x + 2} + \frac{5}{x^2 - x - 6}$$

5. Perform the following subtraction:
$$\frac{x - 3}{x + 5} - \frac{x - 5}{x + 3}$$

6. Solve the following equation:
$$\frac{3}{x - 2} - \frac{4}{x + 1} = \frac{5}{x^2 - x - 2}$$

Answers

1. $\dfrac{1}{x - 3}$. Excluded values are $x = -1, -2,$ and 3.

2. $\dfrac{x + 7}{x - 4}$

3. $\dfrac{x - 1}{x + 1}$

4. $\dfrac{1}{x - 3}$

5. $\dfrac{16}{(x + 3)(x + 5)}$

6. $x = 6$.

CHAPTER 12

Exponential and Logarithmic Functions

Do I Need to Read This Chapter?

➡ What are exponential and logarithm functions?

➡ What are common and natural logarithms?

➡ How do I work with exponential and logarithm functions using calculators, Minitab, and Excel?

➡ What are the basic properties of logarithms?

➡ What are some of the common mistakes made when working with logarithms?

Exponential Functions

Example 12-1: Everyone likes to watch his or her money grow. Compound interest is the mechanism by which we can grow our money. The compound interest formula is

$$A = P\left(1 + \frac{r}{n}\right)^{nt}$$

where A is the amount present after t years when a principal P dollars is invested at an interest rate r, compounded n times per year. Suppose that you invest $P = \$1000$ for $t = 5$ years at a rate $r = 6\%$ compounded quarterly (that is, $n = 4$). The amount after 5 years is

$$A = 1000\left(1 + \frac{0.06}{4}\right)^{20} = 1000(1.015)^{20}$$

Using a calculator to evaluate $(1.015)^{20}$, we find $(1.015)^{20} = 1.34686$, and $A = \$1346.86$. Table 12-1 shows the amount at the end of years 1 through 30. The quantities in the amount column were obtained by substituting the values $1, 2, \cdots, 30$ for t in the equation

$$A = 1000\left(1 + \frac{0.06}{4}\right)^{4t} = 1000(1.015)^{4t}$$

The compound interest formula is an example of an **exponential function. Functions of the form $y = a^x$, where $a > 0$, $a \neq 1$, and x and y are variables, are called exponential functions.** The number a is called the **base** of the exponential function. Exponential functions are useful in describing real-world phenomena such as the growth of money, the growth of populations, radioactive decay, and so on.

Table 12-1   The Growth of $1000 over 30 Years at 6% Compounded Quarterly

YEAR	AMOUNT	YEAR	AMOUNT	YEAR	AMOUNT
1	1061.36	11	1925.33	21	3492.59
2	1126.49	12	2043.48	22	3706.91
3	1195.62	13	2168.87	23	3934.38
4	1268.99	14	2301.96	24	4175.80
5	1346.86	15	2443.22	25	4432.05
6	1429.50	16	2593.14	26	4704.01
7	1517.22	17	2752.27	27	4992.67
8	1610.32	18	2921.16	28	5299.03
9	1709.14	19	3100.41	29	5624.20
10	1814.02	20	3290.66	30	5969.32

Figure 12-1 is a Minitab plot of the data in Table 12-1

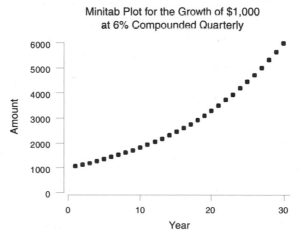

Fig. 12-1

Example 12-2: The median price for a new automobile is $20,000. A median-priced auto is purchased. Suppose it retains 85% of its previous value each year that it remains in use. The value of the automobile y after x years is given by the exponential equation

$$y = 20,000(0.85)^x$$

At the end of 5 years, the value for this auto would be $y = 20,000(0.85)^5$. Using a calculator to evaluate $(0.85)^5$, we find that $y = 20,000(0.4437) = \$8,874$. Table 12-2 shows the value of a median-priced automobile at the end of each year for 15 years.

Table 12-2 Depreciated Values for a Median-Priced Automobile

YEAR	VALUE	YEAR	VALUE	YEAR	VALUE
1	17000.0	6	7543.0	11	3346.9
2	14450.0	7	6411.5	12	2844.8
3	12282.5	8	5449.8	13	2418.1
4	10440.1	9	4632.3	14	2055.4
5	8874.1	10	3937.5	15	1747.1

Figure 12-2 shows a Minitab plot of the data in Table 12-2.

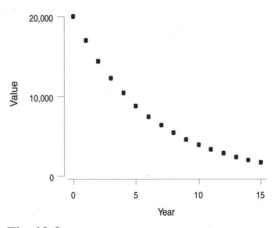

Minitab Plot of the Depreciated Values
for a New Median-Priced Automobile

Fig. 12-2

Quick Tip

When we compare Figs. 12-1 and 12-2, we see that the exponential function in Fig. 12-1 is an example of an increasing function. That is, as the years increase, the amount increases. The exponential function in Fig. 12-2 is an example of a decreasing function. That is, as the years increase, the value of the automobile decreases. This is called depreciation. *When a is greater than 1, the function $f(x) = a^x$ is an increasing function. The greater the value of a, the faster the function increases. When a is between 0 and 1, the function $f(x) = a^x$ is a decreasing function. The greater the value of a, the more slowly the function decreases.*

Logarithm Function

A function that is very closely related to the exponential function is the logarithm function. In fact, the exponential and logarithm functions are inverse functions. **For all positive numbers, a, $a \neq 1$, and all positive numbers x, $y = \log_a x$ means the same as $x = a^y$.** The number a is called the **base** of the logarithm. Note that the **logarithm** of a number is actually an exponent. At first, the logarithm function seems difficult to understand. I have found over many years of teaching algebra that the best way to grasp the definition is to simply practice it over and over.

Example 12-3: Suppose we wish to find $\log_2 8$. What exponent of 2 will give 8? The answer is 3 since $2^3 = 8$. Therefore, we write $\log_2 8 = 3$. Table 12-3 gives several illustrations intended to help you grasp the concept of a logarithm. Read each row of this table carefully and when you finish, you will have a much better understanding of a logarithm. Table 12-3 assures us that by using our knowledge of exponents, we can often find the logarithm of a number in our head. But this is not always the case. Suppose we wish to find $\log_2 7.5$. It is not obvious what exponent of 2 will give 7.5. That is the solution to $2^x = 7.5$ is not clear. We will show you how to find this logarithm in the following sections.

Table 12-3   Practicing the Definition of a Logarithm

FIND	WHAT QUESTION MUST YOU ASK?	ANSWER
$\log_5 25$	What exponent of 5 will give 25?	$\log_5 25 = 2$
$\log_4 64$	What exponent of 4 will give 64?	$\log_4 64 = 3$
$\log_{10} 100$	What exponent of 10 will give 100?	$\log_{10} 100 = 2$
$\log_{12} 144$	What exponent of 12 will give 144?	$\log_{12} 144 = 2$
$\log_c c^3$	What exponent of c will give c^3?	$\log_c c^3 = 3$
$\log_b b$	What exponent of b will give b?	$\log_b b = 1$
$\log_b 1$	What exponent of b will give 1?	$\log_b 1 = 0$
$\log_2 0.25$	What exponent of 2 will give 0.25? $(0.25 = 2^{-2})$	$\log_2 0.25 = -2$
$\log_{10} 0.001$	What exponent of 10 will give 0.001? $(0.001 = 10^{-3})$	$\log_{10} 0.001 = -3$

Common and Natural Logarithms

Although any positive number $a \neq 1$ may be used as the base for logarithms, two bases are commonly used. When the base 10 is used, the logarithms are called **common logarithms.** When the irrational number e that is approximately equal to 2.71828 is used, the logarithms are called **natural logarithms.** The number e is usually first encountered when students take their first calculus course. Common logarithms are written as $\log x$ rather than $\log_{10} x$ and natural logarithms are written as $\ln x$ rather than $\log_e x$.

Example 12-4: Prior to the widespread availability of inexpensive handheld calculators, tables of common and natural logarithms were used to evaluate logarithms. For most scientific calculators, to find the common log of the number 7 for example, you simply enter 7 and then press the **log** key. You will obtain the following:

> Enter 7   followed by log   gives 0.84509804

To obtain the natural log, you enter 7, press the **ln** key, and obtain the following:

> Enter 7   followed by ln   gives 1.945910149

Quick Tip

On many scientific calculators the second or inverse function for log is given as 10^x and the second or inverse function for ln is given as e^x. If you enter the number 0.84509804 and then press second or inv and then log, you obtain the following.

Enter 0.84509804   followed by 2nd   followed by log   gives 7

That is $10^{0.84509804} = 7$.

Similarly, if you enter the number 1.94591049 and then press second or inv and then ln, you obtain

Enter 1.94591049   followed by 2nd   followed by ln   gives 7

That is $e^{1.94591049} = 7$.

Technology and Exponential/ Logarithmic Functions

Many software programs may be used to evaluate exponential and logarithmic functions as well as to graph such functions. Figure 12-3 shows an Excel spreadsheet where certain common and natural logs have been computed. After entering the numbers from 0.5 to 10 in steps of 0.5 in cells A1 to A20, the calculation **=log10(a1)** is entered in C1 and a click and drag is performed to produce the entries shown in column C. Similarly **=ln(a1)** is entered in cell E1 and a click and drag is executed. Note that the same value for the common and natural log of 7 is obtained as with the calculator in Example 12-4.

Figure 12-4 shows the session window as well as the worksheet from Minitab. The commands that entered the numbers in column C1 as well as the common and natural log computations are shown in the session window. The lower half of the window is the worksheet. The values for the logs are shown in the worksheet. Figure 12-5 shows a plot of the common log function produced by MAPLE. The point (10,1) shown on the graph indicates that log 10 = 1.

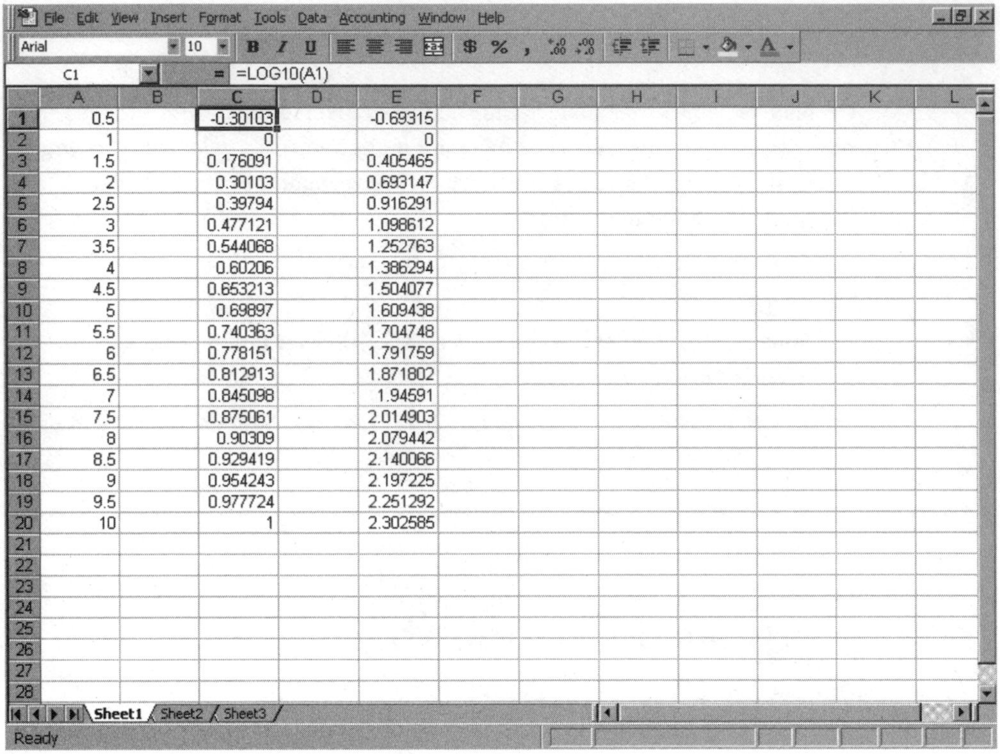

ALGEBRA FOR THE UTTERLY CONFUSED

File Edit View Insert Format Tools Data Accounting Window Help

Arial 10 B I U $ % ,

C1   = =LOG10(A1)

	A	B	C	D	E
1	0.5		-0.30103		-0.69315
2	1		0		0
3	1.5		0.176091		0.405465
4	2		0.30103		0.693147
5	2.5		0.39794		0.916291
6	3		0.477121		1.098612
7	3.5		0.544068		1.252763
8	4		0.60206		1.386294
9	4.5		0.653213		1.504077
10	5		0.69897		1.609438
11	5.5		0.740363		1.704748
12	6		0.778151		1.791759
13	6.5		0.812913		1.871802
14	7		0.845098		1.94591
15	7.5		0.875061		2.014903
16	8		0.90309		2.079442
17	8.5		0.929419		2.140066
18	9		0.954243		2.197225
19	9.5		0.977724		2.251292
20	10		1		2.302585
21					
22					
23					
24					
25					
26					
27					
28					

Sheet1 / Sheet2 / Sheet3

Ready

Fig. 12-3

The author believes strongly that technology should be integrated into the teaching of algebra at all levels. Technology provides strong enhancements to learning algebra.

Properties of Logarithms

Table 12-4 gives four important properties of logarithms. These properties are important in solving many real-world problems that involve logarithms.

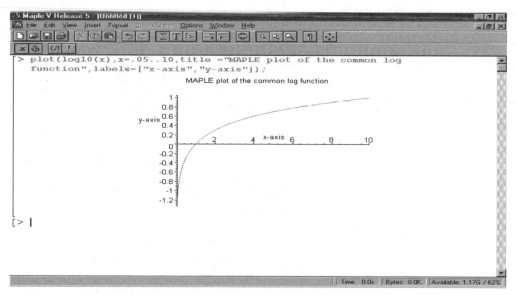

	C1	C2	C3	C4	C5	C6	C7	C8	C9
↓	x		natural		common				
1	0.5		0.60316		0.30103				
2	1.0		0.00000		0.00000				
3	1.5		0.40547		0.17609				
4	2.0		0.69315		0.30103				
5	2.5		0.91629		0.39794				
6	3.0		1.09861		0.47712				
7	3.5		1.25276		0.54407				
8	4.0		1.38629		0.60206				
9	4.5		1.50408		0.65321				
10	5.0		1.60944		0.69897				
11	5.5		1.70475		0.74036				
12	6.0		1.79176		0.77815				

Fig. 12-4

Fig. 12-5

Table 12-4　Properties of Logarithms

PROPERTY	PROPERTY IN WORDS
1. $\log_a (A \cdot B) = \log_a A + \log_a B$	The log of a product is the sum of the logs.
2. $\log_a \left(\dfrac{A}{B} \right) = \log_a A - \log_a B$	The log of a quotient is the difference of the logs.
3. $\log_a A^r = r(\log_a A)$	The log of a number to an exponent is the exponent times the log of the number.
4. $\log_a A = \dfrac{\log_b A}{\log_b a}$	Change of base property.

Example 12-5:　In Example 12-1, $1000 was invested at 6% compounded quarterly. How long will it take for the investment to double? Recall that the amount at any time is given by

$$A = 1000 \left(1 + \frac{0.06}{4} \right)^{4t} = 1000(1.015)^{4t}$$

The answer to the problem is the solution to the equation

$$2000 = 1000(1.015)^{4t}$$

Taking the common log of both sides of this equation, we get

$$\log(2000) = \log[1000(1.015)^{4t}]$$

Letting $A = 1000$ and $B = (1.015)^{4t}$, property 1 in Table 12-4 allows us to reexpress the right-hand side as follows:

$$\log[1000(1.015)^{4t}] = \log 1000 + \log (1.015)^{4t}$$

Using property 3 in Table 12-4 with $A = 1.015$ and $r = 4t$, we can reexpress $\log (1.015)^{4t}$ as follows:

$$\log (1.015)^{4t} = 4t \log (1.015)$$
$$\log 2000 = \log 1000 + 4t \log (1.015)$$

Evaluating the common logs using a calculator, we have

$$3.30103 = 3 + 4t(0.006466)$$

Simplifying, we obtain

$$0.30103 = t(0.025864)$$

$$t = \frac{0.30103}{0.025864} = 11.64 \text{ years}$$

It would take between 11 and 12 years for the investment to double its value.

Danger!

Students often use properties that the logarithm function does not have. Table 12-5 gives properties that are not valid for the logarithm function. *Watch out and do not commit the mistake of using these incorrect properties for logarithms.*

Table 12-5   Watch Out for the Following Incorrect Properties of Logs

DO NOT MAKE THESE MISTAKES!!!
$\log (A + B) \neq \log A + \log B$
$\log (A - B) \neq \log A - \log B$
$\log (A \cdot B) \neq \log A \cdot \log B$
$\log \dfrac{A}{B} \neq \dfrac{\log A}{\log B}$
$\dfrac{\log A}{\log B} \neq \log A - \log B$

Example 12-6: The change of base property of logarithms given in Table 12-4 is often useful and is illustrated as follows. Suppose you need to find the $\log_2 5$. In the equation $\log_a A = \log_b A / \log_b a$, let $a = 2$ and $A = 5$, and choose $b = 10$. Then $\log_2 5$ may be found using common logs as follows:

$$\log_2 5 = \frac{\log 5}{\log 2}$$

Using a calculator we find that the common log of 5 is 0.6989700 and that the common log of 2 is 0.3010299 and that

$$\log_2 5 = \frac{0.6989700}{0.3010299} = 2.3219$$

If b is chosen to equal e, then $\log_2 5$ may be found using natural logs as follows:

$$\log_2 5 = \frac{\ln 5}{\ln 2}$$

Using a calculator we find that the natural log of 5 is 1.6094379 and that the natural log of 2 is 0.6931472 and that

$$\log_2 5 = \frac{1.6094379}{0.6931472} = 2.3219$$

Thus, we see that $\log_2 5$ may be found by using either common or natural logs and the change of base property.

Don't Forget

Functions of the form $y = a^x$, where $a > 0$, $a \neq 1$, and x and y are variables, are called *exponential functions*. The number a is called the *base* of the exponential function.

For all positive numbers a, $a \neq 1$, and all positive numbers x, $y = \log_a x$ means the same as $x = a^y$. The number a is called the base of the *logarithm*.

When the base of the logarithm is 10, the logarithm is called a *common log*. When the base is e, the logarithm is called a *natural log*.

$$\log_a (A \cdot B) = \log_a A + \log_a B$$

$$\log_a \left(\frac{A}{B}\right) = \log_a A - \log_a B$$

$$\log_a A^r = r(\log_a A)$$

$$\log_a A = \frac{\log_b A}{\log_b a}$$

Test Yourself

Questions

1. Evaluate the following logarithms without the use of a calculator or computer software:

   (a) $\log_5 125$

   (b) $\log_{13} 169$

   (c) $\log_2 0.125$

   (d) $\log_7 0.002915452$

   (e) $\log_a a^4$

2. Evaluate the following common and natural logs by the use of a calculator or computer software:

   (a) $\log 175$

   (b) $\log 0.245$

   (c) $\ln 175$

   (d) $\ln 0.245$

3. Use the change of base property to evaluate the following logarithms:

   (a) $\log_5 13$

   (b) $\log_{13} 5$

4. Use the properties of logarithms to express the following as a sum or difference of logarithms:

   (a) $\log \dfrac{x^2 y^3}{z^4}$

   (b) $\log \dfrac{\sqrt{a}\, b^3}{\sqrt[3]{c}}$

5. Use the properties of logarithms to express the following as a single logarithm.

   (a) $2 \log x - 3 \log y + \log z$

   (b) $0.5 \log x + 0.25 \log y - 3 \log z$

Answers

1. (a) 3;   (b) 2;   (c) –3;   (d) –3;   (e) 4

2. (a) 2.24304;   (b) –0.61083;   (c) 5.164786;   (d) –1.40650

3. (a) 1.59369;   (b) 0.62747

4. (a) $2 \log x + 3 \log y - 4 \log z$;   (b) $\dfrac{1}{2} \log a + 3 \log b - \dfrac{1}{3} \log c$

5. (a) $\log \dfrac{x^2 z}{y^3}$;   (b) $\log \dfrac{\sqrt{x}\, \sqrt[4]{y}}{z^3}$

CHAPTER 13

Equations of Second Degree and Their Graphs

Do I Need to Read This Chapter?

➡ What is the standard equation of a circle and how can I investigate a circle using computer software?

➡ How do I find the center and radius of a circle by completing the square?

➡ What is the standard equation of an ellipse and how can I investigate an ellipse using computer software?

➡ How do I put the equation of an ellipse into standard form by completing the square?

➡ What is the standard equation of a hyperbola and how can I investigate a hyperbola using computer software?

➡ How do I put the equation of a hyperbola into standard form by completing the square?

➡ What are the asymptotes of a hyperbola?

Circles

The set of all points that are the same distance from a fixed point, called the **center,** forms a **circle.** Suppose the center is located at a point (h, k) in a rectangular coordinate system and that all the points on the circle are r units from the center. The distance r is called the **radius** of the circle. The equation of the circle is as follows:

$$(x - h)^2 + (y - k)^2 = r^2$$

Example 13-1: Investigate the circle centered at the origin and having radius equal to 1. Because the circle has its center at the origin, $h = 0$ and $k = 0$. And since the radius is equal to 1, $r = 1$. The equation is therefore

$$x^2 + y^2 = 1$$

This circle is often called the **unit circle.** Four of the points located on this circle are $(0, 1), (1, 0), (0, -1)$, and $(-1, 0)$. It is clear that these points are on the circle and satisfy the equation since the following are true: $0^2 + 1^2 = 1, 1^2 + 0^2 = 1, 0^2 + (-1)^2 = 1$, and $(-1)^2 + 0^2 = 1$. Notice also that the points do not satisfy the definition of a function since the points $(0, 1)$ and $(0, -1)$ have the same first coordinate but different second coordinates. The set of points that are on a circle do not satisfy the definition of a function. However, if the equation is solved for y, two separate functions are obtained. Solving for y by subtracting x^2 from both sides and taking the square root we obtain

$$y = \sqrt{1 - x^2} \qquad \text{and} \qquad y = -\sqrt{1 - x^2}$$

When computer algebra software such as MAPLE is used to graph a circle, the two separate functions are used. Figure 13-1 shows the command and the graph for the unit circle.

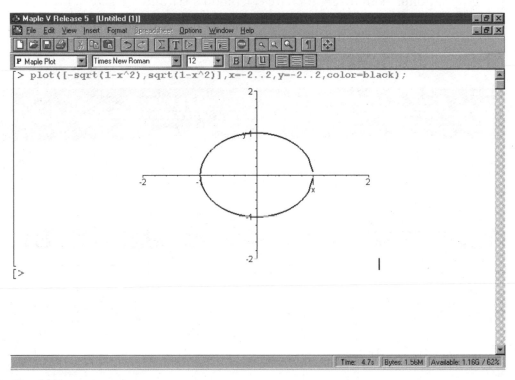

Fig. 13-1

Quick Tip

In Fig. 13-1, note that the upper half of the circle is described by the equation $y = \sqrt{1 - x^2}$ and the lower half of the circle is described by the equation $y = -\sqrt{1 - x^2}$. The upper half represents a function because it satisfies the vertical-line test. That is, any vertical line cuts the graph of the upper half in only one point. Similarly, the lower half of the circle satisfies the vertical-line test.

Example 13-2: Suppose we are interested in finding some of the points that comprise the circle in Fig. 13-1.

Figure 13-2 shows the Minitab computation for some of the points comprising the circle. Columns C1 and C2 give some of the coordinates for points on the upper half of the circle. Columns C1 and C3 give some of the coordinates for points on the lower half of the circle.

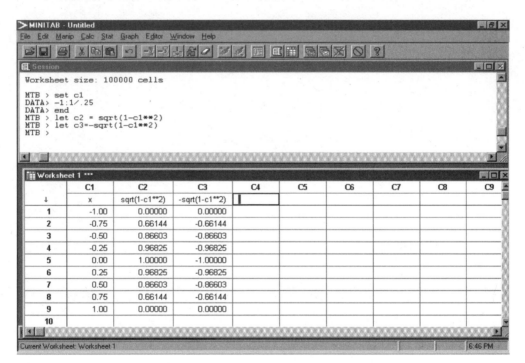

Fig. 13-2

Example 13-3: Investigate the circle centered at $(2, 3)$ and having radius $r = 2$. The equation of this circle is

$$(x - 2)^2 + (y - 3)^2 = 2^2 = 4$$

To obtain an Excel plot, we must solve the equation for y. First, we subtract $(x - 2)^2$ from both sides of the equation to obtain

$$(y - 3)^2 = 4 - (x - 2)^2$$

Taking the square root of both sides of the equation we find

$$y - 3 = \pm\sqrt{4 - (x - 2)^2}$$

Finally, adding 3 to both sides, we find

$$y = 3 \pm\sqrt{4 - (x - 2)^2}$$

The upper half of the circle is given by the equation $y - 3 + \sqrt{4 - (x - 2)^2}$ and the lower half of the circle is given by $y = 3 - \sqrt{4 - (x - 2)^2}$. An Excel plot of the lower half and upper half of the circle is shown in Fig. 13-3. Column A contains the x values from 0 to 4 in steps of 0.1 units. Column B contains the y values for the lower half of the circle and column C contains the y values for the upper half of the circle. From the figure, four points on the circle are

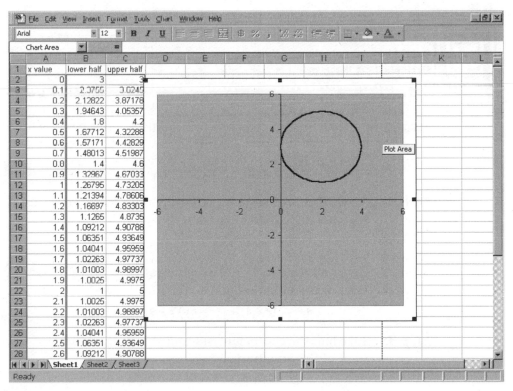

Fig. 13-3

easy to determine from the graph. These points are (2, 1), (4, 3), (2, 5), and (0, 3). These points are located at the top and bottom and on the far right and far left sides of the circle. Note that these points are all 2 units from (2, 3), the center of the circle.

Danger!

When you are given the center and radius of a circle and asked to find the equation of the circle, be careful with the signs. Suppose the center is given to be $(h, k) = (-4, 5)$ and the radius as $r = 3$. The equation is then

$$[x - (-4)]^2 + (y - 5)^2 = 3^2 \quad \text{or} \quad (x + 4)^2 + (y - 5)^2 = 9$$

Pattern

Sometimes the equation of the circle is given in a form that requires you to perform a *completing of the square* to find the center and the radius. Example 13-4 illustrates this technique.

Example 13-4: Find the center and radius for the circle whose equation is

$$x^2 + y^2 - 4x - 6y + 9 = 0$$

Step 1: Rewrite the equation by gathering the x terms and y terms together as follows:

$$(x^2 - 4x\ ) + (y^2 - 6y) = -9$$

Step 2: Take half the coefficient of x and square it and add to both sides of the equation:

$$(x^2 - 4x + 4) + (y^2 - 6y) = -9 + 4$$

Step 3: Take half the coefficient of y and square it and add to both sides of the equation:

$$(x^2 - 4x + 4) + (y^2 - 6y + 9\ ) = -9 + 4 + 9$$

Step 4: You have completed the square for the x terms and the y terms as follows:

$$(x - 2)^2 + (y - 3)^2 = 4$$

Now $(h, k) = (2, 3)$, $r = 2$, and the equation represents a circle with center at $(2, 3)$ and radius 2.

Ellipses

An **ellipse** is defined as the set of points in a rectangular coordinate system the sum of whose distances from two fixed points is a constant. Each fixed point is called a **focus** of the ellipse. The orbits of the planets and some of the satellites are near elliptical. The **standard form for the equation of an ellipse** centered at the origin is

$$\frac{x^2}{a^2} + \frac{y^2}{b^2} = 1$$

where $a > 0$, $b > 0$, and $a \neq b$.

Quick Tip

Just as with circles, the points that comprise an ellipse do not satisfy the definition of a function. However, by solving the equation of the ellipse for y we obtain two equations that are functions. They represent the upper half and lower half of the ellipse. The solution for y proceeds as follows. First subtract the term involving x^2 from both sides.

$$\frac{y^2}{b^2} = 1 - \frac{x^2}{a^2}$$

Multiplying both sides by b^2 and taking the square root of both sides, we get

$$y = \frac{\sqrt{a^2 b^2 - b^2 x^2}}{a} \quad \text{and} \quad -\frac{\sqrt{a^2 b^2 - b^2 x^2}}{a}$$

Short Cuts

When software is used to plot ellipses, the equations for the lower and upper halves of the ellipse are used because both represent functions.

Example 13-5: Graph the ellipse given by the equation $\dfrac{x^2}{4} + \dfrac{y^2}{100} = 1$. The upper half of the ellipse is given by $y = \dfrac{\sqrt{400 - 100x^2}}{2}$ and the lower half is given by $y = -\dfrac{\sqrt{400 - 100x^2}}{2}$. An Excel plot is shown in Fig. 13-4.

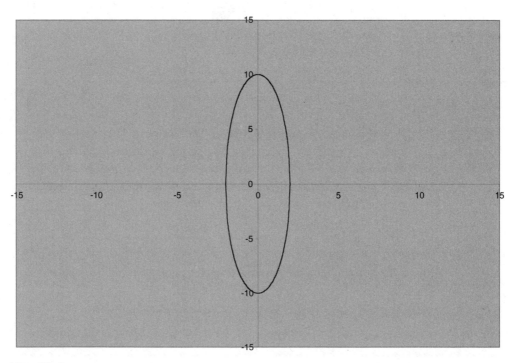

Fig. 13-4

Note that in Fig. 13-4, the y-intercepts are $(0, 10)$ and $(0, -10)$ and the x-intercepts are $(2, 0)$ and $(-2, 0)$. The x-intercepts may be found by letting $y = 0$ in the equation $\dfrac{x^2}{4} + \dfrac{y^2}{100} = 1$ and solving for x. This gives $\dfrac{x^2}{4} + \dfrac{0^2}{100} = 1$ or $x^2 = 4$. Therefore $x = \pm\, 2$. Similarly, if we set $x = 0$ and solve for y, we find that $y = \pm\, 10$.

Pattern

The ellipse that is centered at the origin has the standard equation $\dfrac{x^2}{a^2} + \dfrac{y^2}{b^2} = 1$. The x-intercepts are located at $(a, 0)$ and $(-a, 0)$ and the y-intercepts are located at $(0, b)$ and $(0, -b)$. If the ellipse is centered at (h, k), then the standard equation of the ellipse is

$$\frac{(x - h)^2}{a^2} + \frac{(y - k)^2}{b^2} = 1$$

Example 13-6: Find the standard equation for the ellipse that is centered at $(2, 4)$ and for which $a = 3$ and $b = 5$. Substituting $h = 2$, $k = 4$, $a = 3$, and $b = 5$ in the standard equation, we find

$$\frac{(x - 2)^2}{9} + \frac{(y - 4)^2}{25} = 1$$

If the binomials are squared and the fractions cleared, the equation takes on a different appearance. First, squaring the binomials, we obtain the following:

$$\frac{x^2 - 4x + 4}{9} + \frac{y^2 - 8y + 16}{25} = 1$$

If both sides of this equation are multiplied by 225 to clear the fractions, we obtain:

$$25(x^2 - 4x + 4) + 9(y^2 - 8y + 16) = 225$$

Simplifying further, we obtain:

$$25x^2 - 100x + 100 + 9y^2 - 72y + 144 = 225$$

Finally, after rearranging terms, we obtain

$$25x^2 + 9y^2 - 100x - 72y + 19 = 0$$

Example 13-7: Suppose you were given the equation $25x^2 + 9y^2 - 100x - 72y + 19 = 0$ and were asked to describe its graph. You would need to put it into standard form in order to describe its graph. The steps to put it into standard form are as follows:

Step 1: Rewrite the equation by gathering the x terms and the y terms together as follows:

$$(25x^2 - 100x) + (9y^2 - 72y) = -19$$

Step 2: Factor the left-hand side as follows:

$$25(x^2 - 4x) + 9(y^2 - 8y) = -19$$

Step 3: Complete the square for the x terms and the y terms as we did in Example 13-4.

$$25(x^2 - 4x + 4) + 9(y^2 - 8y + 16) = -19 + 244$$

Note that 244 needed to be added to the right-hand side since completing the square actually added $25(4) = 100$ and $9(16) = 144$ to the left-hand side.

Step 4: You have completed the square for the x terms and the y terms as follows:

$$25(x - 2)^2 + 9(y - 4)^2 = 225$$

Step 5: Finally, divide both sides by 225 to obtain the following equation:

$$\frac{25(x - 2)^2}{225} + \frac{9(y - 4)^2}{225} = \frac{225}{225}$$

Reducing fractions, we obtain:

$$\frac{(x-2)^2}{9} + \frac{(y-4)^2}{25} = 1$$

Comparing this equation to standard form, we see that the ellipse is centered at $(2, 4)$ and that $a = 3$ and $b = 5$.

Hyperbolas

A **hyperbola** is defined to be a set of points in the rectangular coordinate system the difference of whose distances from two fixed points is a constant. The two fixed points are called the **foci** of the hyperbola. There are two different types of hyperbolas. One type opens up and down and the other type opens right and left. First we will consider hyperbolas with center at the origin. The standard equations for such hyperbolas are

$$\frac{x^2}{a^2} - \frac{y^2}{b^2} = 1 \qquad \text{or} \qquad \frac{y^2}{b^2} - \frac{x^2}{a^2} = 1$$

Example 13-8: Investigate the hyperbola $\dfrac{y^2}{16} - \dfrac{x^2}{4} - 1$. First, solve for the two halves of the hyperbola by solving the equation of the hyperbola for y. Solving for y^2 and then taking the square roots we obtain $y = 4\sqrt{1 + \dfrac{x^2}{4}}$ and $y = -4\sqrt{1 + \dfrac{x^2}{4}}$.

These equations were used to compute the y-values shown in columns B and C in the Excel spreadsheet in Fig. 13-5. The Excel plot of the points is shown in the same figure. This hyperbola opens up and down. Note that the y-intercepts are $(0, -4)$ and $(0, 4)$. The upper half of the hyperbola is described by the equation $y = 4\sqrt{1 + \dfrac{x^2}{4}}$ and the lower half of the hyperbola is described by the equation $y = -4\sqrt{1 + \dfrac{x^2}{4}}$.

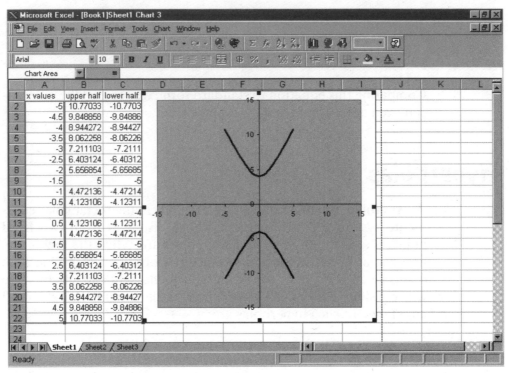

Fig. 13-5

Pattern

The hyperbola $\dfrac{x^2}{a^2} - \dfrac{y^2}{b^2} = 1$ has x-intercepts $(-a, 0)$ and $(a, 0)$. The

hyperbola $\dfrac{y^2}{b^2} - \dfrac{x^2}{a^2} = 1$ has y-intercepts $(0, -b)$ and $(0, b)$. If the

hyperbola has center at (h, k), the standard form for the equations are

$\dfrac{(x - h)^2}{a^2} - \dfrac{(y - k)^2}{b^2} = 1$ and $\dfrac{(y - k)^2}{b^2} - \dfrac{(x - h)^2}{a^2} = 1$.

Asymptotes for Hyperbolas

Hyperbolas centered at the origin have equations of the form $\dfrac{x^2}{a^2} - \dfrac{y^2}{b^2} = 1$ or $\dfrac{y^2}{b^2} - \dfrac{x^2}{a^2} = 1$. These hyperbolas have asymptotes given by the equations $y = \pm \dfrac{b}{a}\, x$. To understand what an asymptote is consider the following example.

Example 13-9: The hyperbola $\dfrac{y^2}{16} - \dfrac{x^2}{4} = 1$ discussed in Example 13-8 and graphed in Fig. 13-5 is discussed further in this example. Note that $a^2 = 4$ and $b^2 = 16$ and therefore $a = 2$ and $b = 4$. The equations of the asymptotes are $y = \pm \dfrac{4}{2}x$ or $y = \pm\, 2x$. The Excel spreadsheet shown in Fig. 13-6 gives the values for the upper and lower halves of the hyperbola as well as values on the lines $y = 2x$ and $y = -2x$. Note that as the x values increase in the positive direction that the upper half values get closer to the values on the line $y = 2x$. As the x values get smaller in the negative direction, the lower half values get closer to the values on the line $y = -2x$. The values on the hyperbola get closer and closer to the lines $y = \pm\, 2x$, but never touch the lines. We say that the lines are **asymptotes** for the hyperbola. This is also shown in the graph in Fig. 13-6.

Example 13-10: Write the following hyperbola in standard form and give the center and the values for a and b. $4x^2 - y^2 - 8x + 6y - 21 = 0$. The steps to put it into standard form are as follows:

Step 1: Rewrite the equation by gathering the x terms and the y terms together as follows:

$$(4x^2 - 8x) - (y^2 - 6y) = 21$$

Step 2: Factor the left-hand side as follows:

$$4(x^2 - 2x) - (y^2 - 6y) = 21$$

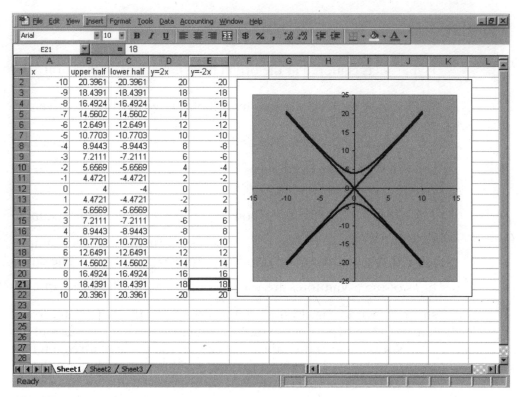

Fig. 13-6

Step 3:   Complete the square for the *x* terms and the *y* terms as we did in Examples 13-4 and 13-7.

$$4(x^2 - 2x + 1) - (y^2 - 6y + 9) = 21 + 4 - 9$$

Step 4:   You have completed the square for the *x* terms and the *y* terms as follows:

$$4(x - 1)^2 - (y - 3)^2 = 16$$

Step 5:   Finally, divide both sides by 16 to obtain the following equation:

$$\frac{4(x - 1)^2}{16} - \frac{(y - 3)^2}{16} = \frac{16}{16}$$

Reducing fractions, we obtain

$$\frac{(x-1)^2}{4} - \frac{(y-3)^2}{16} = 1$$

Comparing this equation to standard form, we see that the hyperbola is centered at $(1, 3)$ and that $a = 2$ and $b = 4$.

Parabolas

Parabolas are also described by equations of the second degree. We have previously discussed parabolas and their properties in Chapter 9.

Don't Forget

A *circle* having center at (h, k) and *radius r* has equation $(x - h)^2 + (y - k)^2 = r^2$.

The *unit circle* is centered at the origin and has radius equal to 1. The equation of the unit circle is $x^2 + y^2 = 1$.

The standard form for the equation of an *ellipse centered at the origin* is $\frac{x^2}{a^2} + \frac{y^2}{b^2} = 1$, where $a > 0$, $b > 0$, and $a \neq b$.

If the *ellipse is centered at* (h, k), then the standard equation of the ellipse is $\frac{(x-h)^2}{a^2} + \frac{(y-k)^2}{b^2} = 1$.

The standard form for the equation of a *hyperbola centered at the origin* is $\frac{x^2}{a^2} - \frac{y^2}{b^2} = 1$ or $\frac{y^2}{b^2} - \frac{x^2}{a^2} = 1$.

If the *hyperbola has center at* (h, k), the standard form for the equations are $\frac{(x-h)^2}{a^2} - \frac{(y-k)^2}{b^2} = 1$ and $\frac{(y-k)^2}{b^2} - \frac{(x-h)^2}{a^2} = 1$.

The *asymptotes for a hyperbola* centered at the origin are $y = \pm \frac{b}{a} x$.

Test Yourself

Questions

1. Find the equation of the circle centered at $(2, -2)$ having radius equal to 2.

2. Find the equation of the ellipse centered at $(-3, -2)$ with $a = 2$ and $b = 4$.

3. Find the equation of the hyperbola that faces right and left with center $(2, -2)$ and having $a = 1$ and $b = 2$. Give the equations for the asymptotes.

4. Write the following equation that represents a circle in standard form and give the center and radius of the circle.

$$x^2 + 6x + y^2 - 10y - 2 = 0$$

5. Write the following equation that represents an ellipse in standard form. Give the center, and the values for a and b.

$$4x^2 - 16x + 9y^2 + 36y + 16 = 0$$

6. Write the following equation that represents a hyperbola in standard form, give the center, the values for a and b, and the equations of the asymptotes.

$$16y^2 - 192y - 9x^2 - 54x + 351 = 0$$

Answers

1. $x^2 + y^2 - 4x + 4y + 4 = 0.$

2. $16x^2 + 4y^2 + 96x + 16y + 96 = 0.$

3. $4x^2 - y^2 - 16x - 4y + 8 = 0$; asymptotes: $y = 2x$ and $y = -2x$.

4. $(x + 3)^2 + (y - 5)^2 = 36$; center $= (-3, 5)$; radius $= 6$.

5. $\dfrac{(x - 2)^2}{9} + \dfrac{(y + 2)^2}{4} = 1$; center $= (2, -2)$; $a = 3, b = 2$.

6. $\dfrac{(y - 6)^2}{9} - (\dfrac{x + 3)^2}{16} = 1$; center $= (-3, 6)$; $a = 4, b = 3$; asymptotes: $y = \pm\dfrac{3}{4}x.$

CHAPTER 14

Sequences and Series

Do I Need to Read This Chapter?

➡ What are a sequence and a series and what is summation notation?

➡ What are the properties of arithmetic sequences and series?

➡ What are the properties of geometric sequences and series?

➡ What is an infinite geometric series, when does it have a sum, and if it has a sum, how can I find the sum?

➡ What are the formulas for summing powers of natural numbers?

Sequences and Series

Example 14-1: When purchasing a home, most Americans secure a loan from a bank or some other financial institution. The loan is repaid as a **sequence** of payments. Consider a loan that is being repaid monthly for 15 years with monthly payments of $750. We represent this sequence of payments as follows. The first **term** in the sequence of payments is represented by a_1, the second term is represented by a_2, and the last term or payment would be a_{180}. We represernt this sequence as follows:

$$a_1 = 750, a_2 = 750, \cdot\cdot\cdot, a_{180} = 750$$

The sum of all the payments is called a **series** and is represented by the following **summation notation:**

$$\sum_{n=1}^{n=180} a_n = a_1 + a_2 + \cdot\cdot\cdot + a_{180} = 750 + 750 + \cdot\cdot\cdot + 750 = 180(750) = \$135,000$$

Example 14-2: In algebra, many different sequences and series are encountered. If $a_1 = 1$, $a_2 = 4$, $a_3 = 9$, and $a_4 = 16$, then an expression for the *n*th term would be $a_n = n^2$. The series consisting of the sum of the first five terms would be

$$\sum_{n=1}^{n=5} a_n = \sum_{n=1}^{n=5} n^2 = 1 + 4 + 9 + 16 + 25 = 55$$

Arithmetic Sequences and Series

Example 14-3: A rod pyramid consists of 40 steel reinforcing rods (used to increase concrete strength) on the bottom row, followed by 39 rods in the row above that row, 38 rods in the next row above, and so forth, until the last row that has only 1 rod. If we let a_1 be the number of rods in the top row, a_2 be the number of rods in the second row from the top, and so forth, the we have the sequence

$$a_1 = 1, a_2 = 2, \cdots, a_{40} = 40$$

The general term is $a_n = n$. Suppose we wish to find the total number of rods in the pyramid. If S_n represents the sum of the first n rows, then the sum in the 40 rows is

$$S_{40} = 1 + 2 + 3 + \cdots + 40$$

This sum may be found by using a calculator to add the 40 numbers or by observing the following mathematical trick. Write the sum forward and backward and then add both sides as follows:

$$S_{40} = 1 + 2 + 3 + \cdots + 40$$
$$\underline{S_{40} = 40 + 39 + 38 + \cdots + 1}$$
$$2S_{40} = 41 + 41 + 41 + \cdots + 41$$
$$2S_{40} = 40(41)$$
$$S_{40} = 20(41) = 820$$

Example 14-3 is an example of an arithmetic sequence. In general, an **arithmetic sequence** is a sequence of numbers in which each term after the first is obtained by adding a constant d to the preceding term. The constant d is called the **common difference.** Suppose a represents the **first term** of the sequence. The terms of the sequence are as follows:

$$a_1 = a$$
$$a_2 = a_1 + d = a + d$$
$$a_3 = a_2 + d = (a + d) + d = a + 2d$$
$$a_4 = a_3 + d = (a + 2d) + d = a + 3d$$
$$\cdots$$
$$a_n = a_{n-1} + d = [a + (n-2)d] + d = a + (n-1)d$$

Summarizing, the nth term of an arithmetic sequence with first term a and common difference d is

$$a_n = a + (n-1)d$$

The sum of an arithmetic sequence is

$$S_n = a + (a+d) + (a+2d) + \cdots + [a + (n-1)d]$$

Using the technique illustrated in Example 14-3, we write the expression for S_n forward and backward and then add term by term as follows:

$$S_n = a + (a+d) + (a+2d) + \cdots + [a + (n-1)d]$$
$$S_n = [a + (n-1)d] + [a + (n-2)d] + \cdots + a$$
$$2S_n = [2a + (n-1)d] + [2a + (n-1)d] + \cdots + [2a + (n-1)d]$$
$$2S_n = n[2a + (n-1)d]$$
$$S_n = \frac{n}{2}[2a + (n-1)d]$$

Example 14-4: Suppose we apply this formula for S_n to Example 14-3 with $a = 1$, $d = 1$, and $n = 40$. Using the formula, we find that $S_{40} = 20[2(1) + (40 - 1)(1)] = 20[2 + 39] = 20[41] = 820$, the same answer obtained in Example 14-3.

Example 14-5: A manufacturer of compact discs sold 100,000 during their first quarter of operation. If sales increased 50,000 per quarter for the next 4 years, what was the total number sold during these 17 quarters?

The sales over the 17 quarters forms an arithmetic sequence with $a = 100{,}000$ and $d = 50{,}000$. The number sold over the 17 quarters is given by S_{17}.

$$S_{17} = \frac{17}{2}[2(100{,}000) + (17 - 1)50{,}000] = 8{,}500{,}000$$

Geometric Sequences and Series

Example 14-6: The formula for compound interest is

$$A = P\,(1 + r)^n$$

where A is the amount present after n years when principal P dollars is invested at an interest rate r, compounded yearly for n years. An **annuity** works as follows. Starting on January 1, 2000, one thousand dollars is deposited in an account *each January* first for 10 years. The account pays 5% interest per year for each of the 10 years. What is the future value of the annuity on January 1, 2010? Table 14-1 shows how the annuity works.

Table 14-1   Annuity

DATE FOR DEPOSIT OF $1000	VALUE OF THIS DEPOSIT ON JANUARY 1, 2010
January 1, 2000	$1000(1.05)^{10} = 1628.89462$
January 1, 2001	$1000(1.05)^9 = 1551.32822$
January 1, 2002	$1000(1.05)^8 = 1477.45544$
January 1, 2003	$1000(1.05)^7 = 1407.10042$
January 1, 2004	$1000(1.05)^6 = 1340.09564$
January 1, 2005	$1000(1.05)^5 = 1276.28156$
January 1, 2006	$1000(1.05)^4 = 1215.50625$
January 1, 2007	$1000(1.05)^3 = 1157.62500$
January 1, 2008	$1000(1.05)^2 = 1102.50000$
January 1, 2009	$1000(1.05)^1 = 1050.00000$
	Total value = 13,206.78715

The values in the second column of Table 14-1 form a geometric series. Listing the values from bottom to top, we have the following:

$a_1 = 1000(1.05)$, $a_2 = 1000(1.05)(1.05)$, a_3

$$= 1000(1.05)(1.05)^2, \cdots, a_{10} = 1000(1.05)(1.05)^9$$

If we let $a = 1000(1.05)$ and $r = 1.05$, then the nth term is given by $a_n = ar^{n-1}$ for $n = 1$ through 10. Such a sequence with first **term a** and a **common ratio r** is called a **geometric sequence.** The general term of a geometric sequence is

$$a_n = ar^{n-1}$$

Now, consider the sum of the annuity. The sum of the 10 terms is

$$S_{10} = 1000(1.05) + 1000(1.05)^2 + \cdots + 1000(1.05)^{10} = 13,206.78715$$

It is possible to derive an expression for the sum of the n terms of a geometric sequence. The sum may be shown to equal

$$S_n = \begin{cases} \dfrac{a(1-r^n)}{1-r}, & r \neq 1 \\ na, & r = 1 \end{cases}$$

Applying this to the annuity described in Example 14-6, we have $r = 1.05$, $a = 1000(1.05) = 1050$, $n = 10$.

$$S_{10} = \frac{1050(1 - 1.05^{10})}{1 - 1.05} = \frac{1050(-0.628895)}{-0.05} = \$13,206.79$$

Example 14-7: BooksPlus.com sold 5000 books in January and the number sold increased by 10% over the previous month for the next 11 months. What is the total number sold over the whole year? Table 14-2 gives the number sold for each of the 12 months as well as the total for the year.

From Table 14-2, we see that the monthly book sales is a geometric sequence with first term $a = 5000$ and common ratio $r = 1.1$. The sales for month n is given by $a_n = 5000(1.1)^{n-1}$. The sum of the sales for n months is given by $S_n = \dfrac{a(1-r^n)}{1-r}$. The yearly sales is given by S_{12}.

Table 14-2   Monthly Book Sales for Bookplus.com

MONTH	BOOKS SOLD THIS MONTH
January	5000
February	$5000(1.1) = 5500$
March	$5000(1.1)^2 = 6050$
April	$5000(1.1)^3 = 6655$
May	$5000(1.1)^4 = 7321$
June	$5000(1.1)^5 = 8053$
July	$5000(1.1)^6 = 8858$
August	$5000(1.1)^7 = 9744$
September	$5000(1.1)^8 = 10,718$
October	$5000(1.1)^9 = 11,790$
November	$5000(1.1)^{10} = 12,969$
December	$5000(1.1)^{11} = 14,266$
	Total sales for the year = 106,924

$$S_{12} = \frac{5000(1 - 1.1^{12})}{1 - 1.1} = \frac{5000(-2.138428)}{-0.1} = 106,921$$

The difference, three books, between the sum given in the table and the sum given by the formula for S_{12} is due to round-off error.

Pattern

Carefully note the difference between an arithmetic sequence and a geometric sequence. For an arithmetic sequence, adding the common difference to any term gives the following term in the sequence. For a geometric sequence, multiplying any term by the common ratio gives the following term in the sequence.

Sum of an Infinite Geometric Sequence

Example 14-8: Repeating decimals are related to the concept of a geometric series. Consider for example the repeating decimal 0.3535353535 · · · This repeating decimal may be written as follows.

$$0.35 + 0.0035 + 0.000035 + 0.00000035 + \cdots$$

This is a geometric series with first term $a = 0.35$ and common ratio $r = 0.01$. However, this series is infinite. That is, it does not have a last term. We call such a geometric series an *infinite geometric series.* If we were to apply the sum formula to the first n terms, we would obtain the following:

$$S_n = \frac{0.35(1 - 0.01^n)}{1 - 0.01} = \frac{0.35(1 - 0.01^n)}{0.99}$$

Note that when n is large, 0.01^n is very small and S_n is close to the fraction 0.35/0.99 or 35/99. In fact, if you convert the fraction 35/99 to its decimal equivalent you get 0.3535353535 · · · This discussion is generalized as follows. Suppose the absolute value of r is less than 1. The sum of the infinite geometric sequence a, ar, ar^2, · · · is equal to

$$S_\infty = \frac{a}{1 - r}$$

When using the formula for S_∞, remember that it is only valid when the absolute value of r is less than 1. Otherwise, the infinite series will *diverge.* That is, the series will not have a finite sum.

Example 14-9: Is it possible to invest a fixed amount of money that will provide yearly income forever? Yes. The financial term for such an investment is called a **perpetuity.** A perpetuity makes use of an infinite geometric series. How many dollars must you invest to accumulate A dollars after n years. The answer is found by solving the equation $A = P(1 + r)^n$ for P. The solution is found by dividing both sides of the equation by the term $(1 + r)^n$. Solving for P, we find

$$P = \left(\frac{1}{1+r}\right)^n A$$

The term $(\frac{1}{1+r})^n$ is called the **discount function.** For example to have $200 after 10 years at a yearly rate of 5%, you must invest the following (note that $\frac{1}{1.05} = 0.95238095$):

$$P = (0.95238095)^{10}(200) = (0.613913254)(200) = \$122.78$$

Table 14-3 shows similar calculations for 1 through 10 years.

Table 14-3   Developing a Perpetuity

YEARS (n)	AMOUNT YOU NEED TO INVEST NOW TO HAVE $200 AFTER n YEARS
1	$(0.95238095)^1(200) = 190.48$
2	$(0.95238095)^2(200) = 181.41$
3	$(0.95238095)^3(200) = 172.77$
4	$(0.95238095)^4(200) = 164.54$
5	$(0.95238095)^5(200) = 156.71$
6	$(0.95238095)^6(200) = 149.24$
7	$(0.95238095)^7(200) = 142.14$
8	$(0.95238095)^8(200) = 135.37$
9	$(0.95238095)^9(200) = 128.92$
10	$(0.95238095)^{10}(200) = 122.78$

If we extend the pattern in Table 14-3, we see that to collect $200 at the end of each year forever, we would need to invest

$(0.95238095)^1(200) + (0.95238095)^2(200)$
$+ (0.95238095)^3(200) + \cdots + (0.95238095)^{10}(200) + \cdots$

We recognize this as an infinite geometric series with first term $a = (0.95238095)^1(200)$ and common ratio $r = 0.95238095$. The sum is

$$S_\infty = \frac{a}{1-r} = \frac{(0.95238095)(200)}{1 - 0.95238095} = \$4000$$

That is, with 5% interest compounded yearly guaranteed forever, $4000 would provide you with $200 every year forever. The general result for such a perpetuity is as follows: If you wish to collect D dollars per year forever and you are guaranteed rate r per year (r expressed in decimal form) then you must invest D/r now. To have a yearly income of $100,000 with 5% interest rate for example, you would need to put $100,000/(0.05) = \$2,000,000$ into the perpetuity.

Summing Powers of Natural Numbers

The following three series are found to be useful in calculus and other mathematics courses.

$$1 + 2 + \cdots + n = \frac{n(n+1)}{2}$$

$$1^2 + 2^2 + \cdots + n^2 = \frac{n(n+1)(2n+1)}{6}$$

$$1^3 + 2^3 + \cdots + n^3 = \frac{n^2(n+1)^2}{4}$$

Example 14-10: Find the sum of the first 1000 natural numbers, the sum of the squares of the first 1000 natural numbers, and the sum of the cubes of the first 1000 natural numbers. Substituting 1000 for n in the above formulas, we find

$$1 + 2 + \cdots + 1000 = \frac{1000(1001)}{2} = 500,500$$

$$1^2 + 2^2 + \cdots + 1000^2 = \frac{1000(1001)(2001)}{6} = 333,833,500$$

$$1^3 + 2^3 + \cdots + 1000^3 = \frac{1000^2(1001)^2}{4} = 250,500,250,000$$

To truly appreciate the usefulness of these formulas, try to verify these results by using a calculator but not the shortcut formulas for the sums.

Don't Forget

An *arithmetic sequence* is a sequence of numbers in which each term after the first is obtained by adding a constant d to the preceding term. The sequence may be written as $a, a + d, a + 2d, \ldots, a + (n - 1)d$. The nth term is $a_n = a + (n - 1)d$. The *sum of an arithmetic sequence* is $S_n = \frac{n}{2}[2a + (n - 1)d]$.

A *geometric sequence* is a sequence of numbers in which each term after the first is obtained by multiplying the preceding term by a constant r. The constant r is called the *common ratio*. The sequence may be written as $a, ar, ar^2, \ldots, ar^{n-1}$. The nth term is $a_n = ar^{n-1}$. The *sum of a geometric series* is $S_n = \frac{a(1 - r^n)}{1 - r}$, $r \neq 1$ and $S_n = na$, $r = 1$.

Provided the absolute value of r is less than 1, an *infinite geometric series* has a sum given by $S_\infty = \frac{a}{1 - r}$.

The following formulas may be used to *sum powers of natural numbers*.

$$1 + 2 + \cdots + n = \frac{n(n + 1)}{2}$$

$$1^2 + 2^2 + \cdots + n^2 = \frac{n(n + 1)(2n + 1)}{6}$$

$$1^3 + 2^3 + \cdots + n^3 = \frac{n^2(n + 1)^2}{4}$$

Questions

Test Yourself

1. Give the values for a, d, a_{10}, and S_{10} for the following arithmetic sequences:

   (a) $13, 15, 17, \ldots$; (b) $1, 7, 13, \ldots$; (c) $5, 3, 1, \ldots$

2. Give the values for a, r, a_{10}, and S_{10} for the following geometric sequences:

   (a) $2, 1, 0.5, \ldots$; (b) $2, 6, 18, \ldots$; (c) $4, -8, 16, \ldots$

3. Starting on January 1, 2000, $5000 is deposited in an account each January first for 15 years. The account pays 6% per year for each of the 15 years. What is the value of the annuity on January 1, 2015?

4. Find S_∞ for each of the following infinite geometric sequences:

   (a) $2, 1, 0.5, \ldots$;   (b) $2, -1, 0.5, \ldots$;   (c) $1, 0.25, 0.0625, \ldots$

5. How much money must be put into a perpetuity now in order to receive $50,000 per year forever if a yearly compound interest rate of 6.5% covering the perpetuity is available?

6. Find the sum, the sum of the squares, and the sum of the cubes for the first 250 natural numbers.

Answers

1. (a) $a = 13, d = 2, a_{10} = 13 + (10 - 1)(2) = 31, S_{10} = 220.$ (b) $a = 1, d = 6, a_{10} = 1 + (10 - 1)(6) = 55, S_{10} = 280.$ (c) $a = 5, d = -2, a_{10} = 5 + (10 - 1)(-2) = -13, S_{10} = -40.$

2. (a) $a = 2, r = 0.5, a_{10} = 2(0.5)^9 = 0.00390625, S_{10} = 3.99609375.$ (b) $a = 2, r = 3, a_{10} = 2(3)^9 = 39{,}366, S_{10} = 59{,}048.$ (c) $a = 4, r = -2, a_{10} = 4(-2)^9 = -2{,}048, S_{10} = -1{,}364.$

3. $123,362.64

4. (a) 4;   (b) 1.333333;   (c) 1.333333

5. $769,230.77

6. 784,375;   5,239,625;   984,390,625

CHAPTER 15

Combinatorics, Binomial Series, and Probability

Do I Need to Read This Chapter?

➡️ What are permutations and combinations and what formulas are used to find them?

➡️ What is a binomial series?

➡️ What are binomial coefficients and how can Pascal's triangle be used to find them?

➡️ What are experiments, sample spaces, and events?

➡️ What are the set operations union, intersection, and complement?

➡️ What is probability?

➡️ How do I compute probabilities of events using the classical definition of probability?

Combinations and Permutations

Often we are interested in the number of ways we can select a subset of objects from a larger set of objects. If the order of the selected objects is of importance to us, then we are interested in the number of possible arrangements or **permutations** that are possible. If the order of the selected objects is not important to us, then we are interested in the number of possible selections or **combinations.** Example 15-1 illustrates the difference in these two concepts.

Example 15-1: Suppose we select two aces from the four aces in a deck of cards. Table 15-1 shows the 12 ordered pairs that are possible. If the order within a pair is not important, then the number of selections possible is 6. Note that the 12 arrangements are divided into 6 pairs. The cards within each pair are the same two cards, just listed in different order. There are 12 permutations and 6 combinations possible when 2 of the 4 aces are selected.

Table 15-1   All 12 Possible Ordered Pairs from the Four Aces

A♣	A♦	A♣	A♥	A♣	A♠
A♦	A♣	A♥	A♣	A♠	A♣
A♥	A♦	A♦	A♠	A♥	A♠
A♦	A♥	A♠	A♦	A♠	A♥

Example 15-2: Suppose we select three aces from the four aces in a deck of cards. Table 15-2 shows the 24 ordered 3-tuples that are possible. If the order within a 3-tuple is not important, then the number of selections possible is 4. There are 4 sets of six 3-tuples each, but the six 3-tuples within each set represent the same basic selection. There are 24 permutations and 4 combinations possible when 3 of the 4 aces are selected.

Table 15-2   All Possible Ordered Triples from the Four Aces

A♣	A♦	A♥	A♣	A♥	A♠
A♣	A♥	A♦	A♣	A♠	A♥
A♥	A♣	A♦	A♥	A♣	A♠
A♥	A♦	A♣	A♥	A♠	A♣
A♦	A♣	A♥	A♠	A♥	A♣
A♦	A♥	A♣	A♠	A♣	A♥
A♣	A♦	A♠	A♦	A♠	A♥
A♣	A♠	A♦	A♦	A♥	A♠
A♦	A♣	A♠	A♠	A♦	A♥
A♦	A♠	A♣	A♠	A♥	A♦
A♠	A♣	A♦	A♥	A♦	A♠
A♠	A♦	A♣	A♥	A♠	A♦

Quick Tip

Remember! When counting the number of ways you may choose a subset of items from a larger set of items, if the order of selection is an important consideration, then you are interested in arrangements or permutations. When the order of selection is not an important consideration, then you are interested in the number of selections or combinations.

Pattern

When n objects are selected from N distinct objects, the number of permutations possible is represented by the symbol $P(N, n)$ and the number of combinations possible is represented by the symbol $C(N, n)$. The following formulas may be used to evaluate the number of possible permutations and combinations:

$$P(N,n) = \frac{N!}{(N-n)!} \quad \text{and} \quad C(N,n) = \frac{N!}{n!(N-n)!}$$

The product of the natural numbers from n to 1, that is, $n(n-1)(n-2)$ $\cdots 1$, is called n *factorial* and is represented by the symbol $n!$.

Example 15-3: Suppose we use these formulas rather than enumeration to work Examples 15-1 and 15-2. With $N = 4$(the number of aces in the deck) and $n = 2$(the number of aces we choose), we find the number of permutations and combinations to be

$$P(4,2) = \frac{4!}{(4-2)!} = \frac{4(3)(2)(1)}{2(1)} = 12$$

and

$$C(4,2) = \frac{4!}{2!(4-2)!} = \frac{4(3)(2)(1)}{2(1)2(1)} = 6$$

With $N = 4$ and $n = 3$, we find

$$P(4,3) = \frac{4!}{(4-3)!} = \frac{4(3)(2)(1)}{1} = 24$$

and

$$C(4,3) = \frac{4!}{3!(4-3)!} = \frac{4(3)(2)(1)}{3(2)(1)(1)} = 4$$

These are the same answers that we found when we listed all the possibilities for Examples 15-1 and 15-2.

Example 15-4: A five-card poker hand is dealt from a deck of 52 cards. How many different hands are possible. The order of the cards in a poker hand is not important when playing poker. Therefore the answer is found by finding the number of combinations possible when 5 cards are selected from 52. The answer is

$$C(52,5) = \frac{52!}{5!47!} = \frac{52(51)(50)(49)(48)47!}{120(47!)} = \frac{52(51)(50)(49)(48)}{120}$$

$$= \frac{311{,}875{,}200}{120} = 2{,}598{,}960$$

Short Cuts

Note in Example 15-4 that expressing 52! as 52(51)(50)(49)(48)47! allows you to divide out the term 47! that appears in the numerator and denominator.

Danger!

When using the factorial button on a calculator (*x*!), be aware that numbers larger than 69! will result in an error. The factorial function cannot be used for numbers larger than 69!

Example 15-5: In the Pick5 lottery, you select 5 numbers from the numbers 1 through 30. How many ways can you select 5 numbers from 30? The order of selection is not important in this lottery. The answer is given by

$$C(30,5) = \frac{30!}{5!25!} = \frac{30(29)(28)(27)(26)(25!)}{5!25!}$$

$$= \frac{30(29)(28)(27)(26)}{120} = 142{,}506$$

Example 15-6: During the later part of 1999, the following individuals were listed as the ones Republicans preferred as their presidential nominees: George W. Bush, John McCain, Steve Forbes, Orrin Hatch, Alan Keyes, and Gary Bauer. If the president/vice president slate came from this group, how many such slates are possible? The number of possible slates is

$$P(6,2) = \frac{6!}{(6-2)!} = \frac{6!}{4!} = 30$$

All the possible slates are listed in Table 15-3. The order of the pairs of individuals is clearly important in this example.

Table 15-3   Possible President/Vice President Slates in Late 1999

Bush	Bush	Bush	Bush	Bush
McCain	Forbes	Hatch	Keyes	Bauer
McCain	McCain	McCain	McCain	McCain
Bush	Forbes	Hatch	Keyes	Bauer
Forbes	Forbes	Forbes	Forbes	Forbes
Bush	McCain	Hatch	Keyes	Bauer
Hatch	Hatch	Hatch	Hatch	Hatch
Bush	McCain	Forbes	Keyes	Bauer
Keyes	Keyes	Keyes	Keyes	Keyes
Bush	McCain	Forbes	Hatch	Bauer
Bauer	Bauer	Bauer	Bauer	Bauer
Bush	McCain	Forbes	Hatch	Keyes

Binomial Series

The goal of this section is to give a systematic method for expanding the binomial $(a + b)$ to the nth power, that is, $(a + b)^n$. The **binomial theorem** states that the expansion for $(a + b)^n$ is

$$(a + b)^n = C(n, 0)a^n + C(n, 1)a^{n-1}b + C(n, 2)a^{n-2}b^2 + \ldots + C(n, n)b^n$$

Table 15-4 gives expansions for $(a + b)^n$ for $n = 0$ through 5. The **binomial coefficients,** shown in bold in the table, form a pattern called **Pascal's triangle.** Any row after the second will have a 1 at the beginning and the end of the row. Furthermore, the coefficients between the ones at the beginning and the end of the row may be found by adding the two coefficients immediately above.

Table 15-4 Binomial Expansions for $n = 0$ through $n = 5$

$(a+b)^n$	1
$(a+b)^1$	$1a + 1b$
$(a+b)^2$	$1a^2 + 2ab + 1b^2$
$(a+b)^3$	$1a^3 + 3a^2b + 3ab^2 + 1b^3$
$(a+b)^4$	$1a^4 + 4a^3b + 6a^2b^2 + 4ab^3 + 1b^4$
$(a+b)^5$	$1a^5 + 5a^4b + 10a^3b^2 + 10a^2b^3 + 5ab^4 + 1b^5$

The row for $(a + b)^6$ would be formed as follows. First write down the coefficients for $(a + b)^5$.

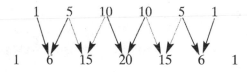

The 6 is found by adding 1 and 5, the 15 is found by adding the 5 and the 10, and so forth.

The row corresponding to $(a + b)^7$ would be found as follows.

$$\begin{array}{ccccccc} 1 & 6 & 15 & 20 & 15 & 6 & 1 \\ 1 & 7 & 21 & 35 & 35 & 21 & 7 & 1 \end{array}$$

It is also of interest to note for the row corresponding to $(a + b)^3$ in Table 15-4 that

$$C(3, 0) = 1 \quad C(3, 1) = 3 \quad C(3, 2) = 3 \quad C(3, 3) = 1$$

A similar relationship holds for any row of Pascal's triangle.

Pattern

The following relationship holds for any set of binomial coefficients:

$$C(n, 0) + C(n, 1) + C(n, 2) + \ldots + C(n, n) = 2^n$$

Example 15-7: The binomial coefficients for the expansion of $(a + b)^7$ are

$$C(7, 0) = 1, C(7, 1) = 7, C(7, 2) = 21, C(7, 3) = 35, C(7, 4)$$
$$= 35, C(7, 5) = 21, C(7, 6) = 7, C(7, 7) = 1$$

The sum of these binomial coefficients is

$$1 + 7 + 21 + 35 + 35 + 21 + 7 + 1 = 128$$

Note that $2^7 = 128$.

Example 15-8: Use the binomial expansion for $(a + b)^4$ to expand $(2x^2 - 4y)^4$. The binomial expansion for $(a + b)^4$ is

$$(a + b)^4 = 1a^4 + 4a^3b + 6a^2b^2 + 4ab^3 + 1b^4$$

To expand $(2x^2 - 4y)^4$ we let $a = 2x^2$ and $b = -4y$ in the expansion for $(a + b)^4$. Applying the laws of exponents, we find that $a^2 = 4x^4, a^3 = 8x^6$, and $a^4 = 16x^8$. Similarly, we find that $b^2 = 16y^2, b^3 = -64y^3$, and $b^4 = 256y^4$. Now, by carefully putting the various parts of the puzzle together, we get:

$$(2x^2 - 4y)^4 = 16x^8 + 4(8x^6)(-4y) + 6(4x^4)(16y^2) + 4(2x^2)(-64y^3) + 256y^4$$
$$(2x^2 - 4y)^4 = 16x^8 - 128x^6y + 384x^4y^2 - 512x^2y^3 + 256y^4$$

To help students get a feel for an expansion such as this, I suggest that they try putting in some values for x and y. For example if $x = 1$ and $y = 1$, the left-hand side of the equation becomes $(-2)^4 = 16$. The right-hand side of the equation becomes $16 - 128 + 384 - 512 + 256 = 16$ and we obtain the same result.

Experiments, Sample Spaces, and Events

Probability is a concept that many of us are familiar with because of its use in everyday conversation. Weather forecasts often are stated with a certain probability associated with precipitation. We may read that it is highly likely (high probability) that the president will veto a certain piece of legislation. We may hear on a news program that it is unlikely (low probability) that the chairman of the federal reserve will raise interest rates. This section will introduce some basic definitions and concepts needed to define probability.

An **experiment** is defined as any operation or procedure whose **outcomes** cannot be predicted with certainty. The **sample space** is defined as the set of all possible outcomes for an experiment. An **event** is any subset of the sample space. These definitions will now be illustrated with some examples.

Example 15-9: For the experiment of rolling a die and observing the number on the side that faces up, give the sample space *S*. The letter *E* represents the event that the side facing up is a prime number. *S* and *E* can be visualized as follows:

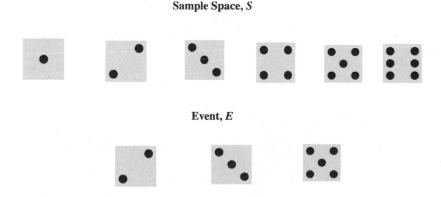

Sample Space, *S*

Event, *E*

Example 15-10: Three coins are flipped and the turned-up side is noted (head or tail). The sample space is given in Fig. 15-1. The sample space is represented by a **tree diagram.** The **branches** of the tree are formed by connecting the side that turned up for each coin by an arrow. The branches of the tree are the potential outcomes for the experiment. The event that all three coins show the same side is represented by the letter *A* and *A* = {*HHH, TTT*}. The event that at least two heads occur in the three tosses is represented by the letter *B* and *B* = {*HHT, HTH, THH, HHH*}.

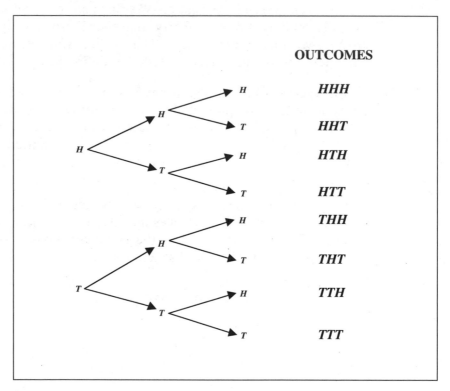

Fig. 15-1

Example 15-11: A pair of dice are rolled. The Excel plot shown in Fig. 15-2 illustrates how the sample space may be represented as a set of rectangular coordinates. The point (1, 1) represents the outcome that a 1 appeared on each of the dice. The point (2, 4) represents the outcome that a 2 turned up on die 1 and a 4 turned up on die 2, and so forth. Suppose A represents the event that the sum of the dice equals 7 and B represents the event that a 3 appeared on die 1. The event A is composed of the 6 outcomes $(1, 6), (2, 5), (3, 4), (4, 3), (5, 2),$ and $(6, 1)$, since the sum equals 7 for each of these outcomes. That is, we may express A as follows:

$$A = \{(1, 6), (2, 5), (3, 4), (4, 3), (5, 2), (6, 1)\}$$

Similarly B may be expressed as follows:

$$B = \{(3, 1), (3, 2), (3, 3), (3, 4), (3, 5), (3, 6)\}$$

Note that for each point in B, a 3 occurred on die 1.

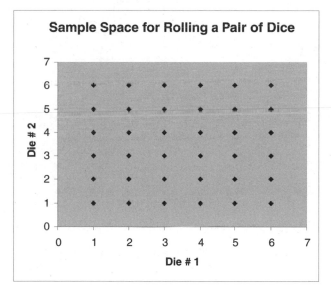

Fig. 15-2

Combining Events

Because events are sets, set operations may be performed on events. The **union** of events A and B consists of those outcomes that belong to *A*, *B*, or both *A* and *B* and it is denoted by $A \cup B$. The **intersection** of events *A* and *B* consists of those outcomes that belong to both *A* and *B* and it is denoted by $A \cap B$. The **complement** of *A* consists of all the points in the sample space not in *A* and is denoted by A^c.

Example 15-12: An experiment consists of selecting 1 card from a standard deck of 52 cards. The sample space *S* may be represented as illustrated in Fig. 15-3. The gray portion of the figure represents the event *A* that a face card is drawn.

A♣	2♣	3♣	4♣	5♣	6♣	7♣	8♣	9♣	10♣	J♣	Q♣	K♣
A♦	2♦	3♦	4♦	5♦	6♦	7♦	8♦	9♦	10♦	J♦	Q♦	K♦
A♥	2♥	3♥	4♥	5♥	6♥	7♥	8♥	9♥	10♥	J♥	Q♥	K♥
A♠	2♠	3♠	4♠	5♠	6♠	7♠	8♠	9♠	10♠	J♠	Q♠	K♠

Fig. 15-3

The event *B* that a club is drawn is the gray portion in Fig. 15-4.

A♣	2♣	3♣	4♣	5♣	6♣	7♣	8♣	9♣	10♣	J♣	Q♣	K♣
A♦	2♦	3♦	4♦	5♦	6♦	7♦	8♦	9♦	10♦	J♦	Q♦	K♦
A♥	2♥	3♥	4♥	5♥	6♥	7♥	8♥	9♥	10♥	J♥	Q♥	K♥
A♠	2♠	3♠	4♠	5♠	6♠	7♠	8♠	9♠	10♠	J♠	Q♠	K♠

Fig. 15-4

The event $A \cup B$ that a face card or a club was drawn is the gray portion in Fig. 15-5.

A♣	2♣	3♣	4♣	5♣	6♣	7♣	8♣	9♣	10♣	J♣	Q♣	K♣
A♦	2♦	3♦	4♦	5♦	6♦	7♦	8♦	9♦	10♦	J♦	Q♦	K♦
A♥	2♥	3♥	4♥	5♥	6♥	7♥	8♥	9♥	10♥	J♥	Q♥	K♥
A♠	2♠	3♠	4♠	5♠	6♠	7♠	8♠	9♠	10♠	J♠	Q♠	K♠

Fig. 15-5

The event $A \cap B$ that a face card and a club are drawn is shaded in gray in Fig. 15-6.

A♣	2♣	3♣	4♣	5♣	6♣	7♣	8♣	9♣	10♣	J♣	Q♣	K♣
A♦	2♦	3♦	4♦	5♦	6♦	7♦	8♦	9♦	10♦	J♦	Q♦	K♦
A♥	2♥	3♥	4♥	5♥	6♥	7♥	8♥	9♥	10♥	J♥	Q♥	K♥
A♠	2♠	3♠	4♠	5♠	6♠	7♠	8♠	9♠	10♠	J♠	Q♠	K♠

Fig. 15-6

The event A^c, that the card drawn is not a face card is the gray portion in Fig. 15-7.

A♣	2♣	3♣	4♣	5♣	6♣	7♣	8♣	9♣	10♣	J♣	Q♣	K♣
A♦	2♦	3♦	4♦	5♦	6♦	7♦	8♦	9♦	10♦	J♦	Q♦	K♦
A♥	2♥	3♥	4♥	5♥	6♥	7♥	8♥	9♥	10♥	J♥	Q♥	K♥
A♠	2♠	3♠	4♠	5♠	6♠	7♠	8♠	9♠	10♠	J♠	Q♠	K♠

Fig. 15-7

Basic Definitions of Probability

The **probability** of an event associated with an experiment is a measure of the likelihood that the event will occur when the experiment is performed. If E represents an event associated with some experiment, we use the symbol $P(E)$ to represent the probability of the event E. When we encounter an event that cannot occur, we refer to such an event as an **impossible event.** Some examples of impossible events are as follows: rolling a pair of dice and getting a sum equal to 13, tossing two coins and observing that three heads occur, and selecting a card from a standard deck and observing the number 11 on the card. The Greek letter ϕ is often used to represent an impossible event. Because an impossible event cannot occur, we require the probability associated with an impossible event to be zero. We sometimes take this as an axiom for a probability measure.

$$P(\phi) = 0$$

Because the sample space S contains all possible outcomes for an experiment, we are certain that one of the outcomes in S will occur when we perform the experiment. Because we are certain that some outcome in S will occur, we require that the probability associated with S equal 1.

$$P(S) = 1$$

Most events associated with an experiment are neither impossible nor certain. Therefore most events will have a probability between 0 and 1. For any event E associated with an experiment, we require the following to be true (the equal signs are included since E might be either the impossible event or S):

$$0 \le P(E) \le 1$$

There are many definitions of probability. We shall consider three commonly used definitions. They are sometimes referred to as the classical definition, the relative frequency definition, and the subjective definition.

The **classical definition of probability** assumes a finite number of equally likely outcomes for an experiment. This definition will be discussed in more detail in the next section.

The **relative frequency definition of probability** does not assume equally likely outcomes but assumes that the experiment may be repeated any number of times. If an experiment is conducted n times and the event E occurs $n(E)$ times in the n repetitions of the experiment, then the probability assigned to E is given by the following:

$$P(E) = n(E)/n$$

The primary shortcoming of the relative frequency definition of probability is that different answers are often obtained for the probability of an event. Because of this, we often refer to the obtained probability as the approximate probability of an event.

Example 15-13: A bent coin is tossed 50 times and a head turns up on 32 of the tosses. What probability would the relative frequency definition assign to the event that a head appears on a single toss of the coin?

The experiment consists of tossing the coin once. We are interested in the event that a head turns up when the coin is tossed. In this example $n = 50$ and $n(E) = 32$. The probability of a head appearing on a single toss is approximately $\frac{32}{50} = .64$.

Example 15-14: In a national survey of 5000 households having a child under 1 year of age, it was found that 4050 of the children under 1 year of age in the survey were constrained by a safety belt when traveling in a car. On the basis of this survey and using the relative frequency definition of probability, what is the probability that a child under 1 year of age will be constrained by a safety belt when traveling by car?

The experiment consisted of observing whether a child under 1 year of age and traveling in a car was constrained by a safety belt. Of the 5000 observed, 4050 were constrained by a safety belt and the probability is approximately $\frac{4050}{5000} = .81$.

The **subjective definition of probability** may be appropriate when neither the classical definition nor the relative frequency definition is appropriate. The subjective definition of probability utilizes experience, intuition, brainstorming sessions, and any available information to assign probabilities.

Example 15-15: An oncologist tells a patient that there is a 90% probability that cancer surgery will be successful. The doctor is assigning a subjective probability of .90 to the event that cancer surgery will be successful for this patient. A military leader states that there is a 10% chance that nuclear weapons will be used in a war within the next 5 years. The military leader is assigning a subjective probability of .10 to the event that nuclear weapons will be used in this conflict. Notice in both of these scenarios that neither the classical nor the relative frequency definitions of probability are appropriate.

Finite Probability Spaces with Equally Likely Outcomes

A **finite probability space** consists of a sample space with a finite number of outcomes. That is, $S = \{O_1, O_2, \cdots, O_k\}$ and each outcome O_i has associated with it a nonnegative number P_i called the **probability of the outcome.** The sum of the P_i is equal to 1. If the k outcomes in S are equally likely, then each P_i will equal $1/k$.

The **classical definition of probability** assumes a finite probability space with equally likely outcomes. When we have a finite probability space with equally likely outcomes, then the probability of any event E is found as follows:

$$P(E) = \frac{\text{number of outcomes in } E}{\text{number of outcomes in } S}$$

Solving Probability Problems Having Equally Likely Outcomes

Example 15-16: The balanced token shown below is tossed twice. The side shown on the left is called tails and the side shown on the right is called heads.

Sample Space

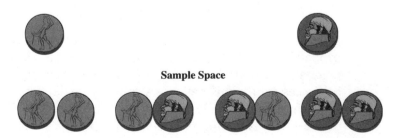

A symbolic representation of the sample space might be as follows.

$$S = \{TT,\ TH,\ HT,\ HH\}$$

Each of the four outcomes is equally likely because the token is balanced.

Let E be the event that at least one tail occurred.

Event E

Or, we may express E as $E = \{TT,\ TH,\ HT\}$ and $P(E) = \dfrac{\text{number of outcomes in } E}{\text{number of outcomes in } S} = \dfrac{3}{4} = .75$. The event that the same side turned up on both tosses is composed of HH and TT and has probability of .50.

Example 15-17: In the game of poker, a royal flush consists of the ace, king, queen, jack, and 10 of the same suit (hearts, diamonds, clubs, or spades). Table 15-5 shows the four possible royal flushes in rows 1 through 4. In Example 15-4, we found that there are 2,598,960 possible hands when 5 cards are dealt from a deck of 52. Four of the 2,598,960 hands constitute a royal flush.

Table 15-5   Royal Flush in the Game of Poker

Clubs royal flush	10♣	J♣	Q♣	K♣	A♣
Diamonds royal flush	10♦	J♦	Q♦	K♦	A♦
Hearts royal flush	10♥	J♥	Q♥	K♥	A♥
Spades royal flush	10♠	J♠	Q♠	K♠	A♠

Using the classical definition of probability, we find the probability of the event E that a royal flush is dealt is

$$P(E) = \frac{\text{number of outcomes in } E}{\text{number of outcomes in } S} = \frac{4}{2,598,960} = 0.000001539$$

Example 15-18: In the Pick5 lottery, 5 numbers are selected from the numbers 1 through 30. In Example 15-5, we found that there are 142,506 ways of selecting 5 numbers from the 30. The probability that you select the 5 winning numbers is therefore equal to 1 divided by 142,506 or 0.000007017.

Example 15-19: Figure 15-2 shows the 36 equally likely outcomes possible when a pair of dice is tossed. In Example 15-11, we found that the outcomes in *A* correspond to rolling a seven with the dice.

$$A = \{(1, 6), (2, 5), (3, 4), (4, 3), (5, 2), (6, 1)\}$$

The probability of rolling a seven is therefore equal to 6 divided by 36 or 0.167.

Don't Forget

When selecting *n* items from a set consisting of *N* items, the number of different arrangements or *permutations* of the *n* items is denoted by $P(N, n)$. The number of selections is denoted by $C(N, n)$. $P(N, n)$ and $C(N, n)$ are given by the following formulas, where $n! = n(n - 1)(n - 2) \cdots 1$.

$$P(N,n) = \frac{N!}{(N - n)!} \quad \text{and} \quad C(N,n) = \frac{N!}{n!(N - n)!}$$

The *expansion for the binomial* $(a + b)^n$ is as follows.

$$(a + b)^n = C(n, 0)a^n + C(n, 1)a^{n-1}b + C(n, 2)a^{n-2}b^2 + \ldots + C(n, n)b^n$$

The first six rows of *Pascal's triangle* are as follows.

$$1$$
$$1 \quad 1$$
$$1 \quad 2 \quad 1$$
$$1 \quad 3 \quad 3 \quad 1$$
$$1 \quad 4 \quad 6 \quad 4 \quad 1$$
$$1 \quad 5 \quad 10 \quad 10 \quad 5 \quad 1$$

The following relationship holds for any set of *binomial coefficients.*

$$C(n, 0) + C(n, 1) + C(n, 2) + \ldots + C(n, n) = 2^n$$

An *experiment* is defined as any operation or procedure whose outcomes cannot be predicted with certainty. The *sample space* is defined as the set of all possible outcomes for an experiment. An *event* is any subset of the sample space.

The *union* of events *A* and *B* consists of those outcomes that belong to *A*, *B*, or both *A* and *B* and it is denoted by $A \cup B$. The *intersection* of events *A* and *B* consists of those outcomes that belong to both *A* and *B* and it is denoted by $A \cap B$. The *complement* of *A* consists of all the points in the sample space not in *A* and is denoted by A^c.

The *classical definition of probability* states that if an experiment can result in *n* equally likely outcomes, and event *E* consists of *k* of these equally likely outcomes, then the probability of *E* occurring is *k* divided by *n*.

The *relative frequency definition of probability* does not assume equally likely outcomes but assumes that the experiment may be repeated any number of times. If an experiment is conducted *n* times and the event *E* occurs *n(E)* times in the *n* repetitions of the experiment, then the probability assigned to *E* is *n(E)* divided by *n*.

The *subjective definition of probability* may be appropriate when neither the classical definition nor the relative frequency definition is appropriate. The subjective definition of probability utilizes experience, intuition, brainstorming sessions, and any available information to assign probabilities.

Test Yourself

Questions

1. If two letters are selected from the letters *A, B,* and *C,* give the possible permutations and the possible combinations.

2. If five of the state governors are chosen to form an advisory committee, how many such committees are possible?

3. The central division of the National Baseball League is made up of the following teams: Houston, Cincinnati, Pittsburgh, St. Louis, Milwaukee, and Chicago. If all six teams have different won-lost records, how many different arrangements are possible for the top-three teams at the end of the season?

4. Use Pascal's triangle to construct the binomial coefficients for the expansion of $(a + b)^8$. Verify that the sum of these coefficients is equal to 2^8.

5. Expand $(x^2 - 3y)^5$.

6. When a die is tossed, what is the probability that the side that faces up is a prime number? (See Example 15-9.)

7. When three coins are tossed, what is the probability that (a) all three coins show the same side? (b) at least two heads occur? (See Example 15-10.)

8. A pair of dice is tossed. What is the probability that a 3 appears on die 1? (See Example 15-11.)

9. If a single card is drawn from a well-shuffled deck, find the probability of the following. (See Example 15-12.)
   (a) A face card is drawn.
   (b) A club is drawn.
   (c) A face card or a club is drawn.
   (d) A face card and a club are drawn.
   (e) A nonface card is drawn.

10. If three dice are tossed, there are 216 possible outcomes in the sample space. What is the probability that the same number of spots appear on all three dice?

Answers

1. Permutations: *AB, BA, AC, CA, BC,* and *CB*
   Combinations: *A* and *B, A* and *C,* and *B* and *C*
2. $C(50, 5) = 2,118,760.$
3. $P(6, 3) = 120.$
4. $1 + 8 + 28 + 56 + 70 + 56 + 28 + 8 + 1 = 256;\ 2^8 = 256.$

5. $x^{10} - 15x^8y + 90x^6y^2 - 270x^4y^3 + 405x^2y^4 - 243y^5$

6. .5

7. (a) .25; (b) .5

8. .167

9. (a) $\dfrac{12}{52}$; (b) $\dfrac{13}{52}$; (c) $\dfrac{22}{52}$; (d) $\dfrac{3}{52}$; (e) $\dfrac{40}{52}$

10. $\dfrac{6}{216}$

Index

About the Author

Larry J. Stephens is currently Professor of Mathematics at the University of Nebraska at Omaha and has been on the faculty since 1974. He received his bachelor's degree from Memphis State University in mathematics, his master's degree from the University of Arizona in mathematics, and his Ph.D. degree from Oklahoma State University in statistics. He has over 25 years of experience teaching mathematics and statistics. He has taught at the University of Arizona, Christian Brothers College, Gonzaga University, Oklahoma State University, the University of Nebraska at Kearney, and the University of Nebraska at Omaha. He has worked for NASA, Livermore Radiation Laboratory, and Los Alamos Laboratory. Dr. Stephens consulted with and conducted statistics seminars for the engineering group at the 3M, Valley, Nebraska, plant from 1989 until 1998. He has published numerous computerized test banks to accompany elementary statistics texts. Professor Stephens has over 40 publications in professional journals. He is the author of *Schaum's Outline of Beginning Statistics* and coauthor with Murray Spiegel of *Schaum's Outline of Statistics,* 3rd Ed.